COMMUNICATION
AGAINST CAPITAL

A volume in the series
Cornell Modern Indonesia Project
Edited by Eric Tagliacozzo and Thomas B. Pepinsky

A list of titles in this series is available at cornellpress.cornell.edu.

COMMUNICATION AGAINST CAPITAL

Red Enlightenment at the Dawn
of Indonesia

Rianne Subijanto

SOUTHEAST ASIA PROGRAM PUBLICATIONS

an imprint of

CORNELL UNIVERSITY PRESS

Ithaca and London

Cornell University Press gratefully acknowledges the role of the Association for Asian Studies First Book Subvention Program for its support of this book.

Copyright © 2025 by Cornell University

First published 2025 by Cornell University Press

Library of Congress Cataloging-in-Publication Data
Names: Subijanto, Rianne, 1982– author.
Title: Communication against capital : red enlightenment at the dawn of Indonesia / Rianne Subijanto.
Description: Ithaca : Southeast Asia Program Publications, an imprint of Cornell University Press, 2025. | Series: Cornell Modern Indonesia Project | Includes bibliographical references and index.
Identifiers: LCCN 2024021684 (print) | LCCN 2024021685 (ebook) | ISBN 9781501778650 (hardcover) | ISBN 9781501778667 (paperback) | ISBN 9781501778674 (epub) | ISBN 9781501778681 (pdf)
Subjects: LCSH: Communication in politics—Indonesia—History—20th century. | Anti-imperialist movements—Indonesia—History—20th century. | Government, Resistance to—Indonesia—History—20th century. | Indonesia—Politics and government—1798–1942.
Classification: LCC JA85.2.I5 S83 2025 (print) | LCC JA85.2.I5 (ebook) | DDC 959.8/0223—dc23/eng/20240820
LC record available at https://lccn.loc.gov/2024021684
LC ebook record available at https://lccn.loc.gov/2024021685

Cover photograph: Leden vergadering (member meeting) S.I. Kaliwoengoe at *Sarekat Islam* office on 25 September 1921 were attended by women, children, and men. Sitting to the right behind the man in the center is Ms. Moenasiah. From the Collection Nationaal Museum van Wereldculturen. Coll.no. TM-60009089.

For my late parents,
Mei & Agus

Contents

Acknowledgments

For over a quarter of my life, this book has taken center stage, accompanying me through a bricolage of events, at times distracting me from moments of joy and elation and at others carrying me through those of grief and anguish. I would like to thank individuals and institutions who have given me the support and encouragement that enabled this work to come to fruition.

My interest in the historical study of Marxist thought and politics developed during the years at the University of Colorado Boulder, where I studied under the mentorship of Janice Peck. Through her guidance, she has challenged me to think beyond the trends and encouraged me to expand my research beyond what I could have imagined it to be. I thank her for continuously guiding me to defend humanity and justice through the rigorous pursuit of knowledge, and for holding my hands throughout these years. At CU Boulder, I am also thankful for Andrew Calabrese, Stewart Hoover, Nabil Echchaibi, Carla Jones, Marcia Yonemoto, and Chad Kautzer, whose challenging questions and ideas helped shape a solid foundation for this book.

The research and writing of this book has been funded at various stages by the American Association of University Women International Fellowship, CU Boulder's James R., Ann R. and R. Jane Emerson (McCall) Dissertation Fellowship, Baruch College's WSAS Dean's Faculty Research Support Award, PSC-CUNY Research Award, the Eugene M. Lang Fellowship, the CUNY Faculty Fellowship Publication Program (FFPP), Villanova University's Waterhouse Family Institute for the Study of Communication and Society Research Grant, the Andrew W. Mellon Society of Fellows in Critical Bibliography, and an award from the Association for Asian Studies (AAS) First Book Subvention Program.

My archival research was hosted by the Southeast Asia Program at Cornell University and the Royal Netherlands Institute of Southeast Asian and Caribbean Studies (KITLV) in Leiden, the Netherlands. I am thankful for the resourceful staff and librarians of Carl A. Kroch Library at Cornell University—especially Ben Abel and Jeff Petersen, KITLV Library—in particular Rini Hogewoning and Yayah Siegers, International Institute of Social History—notably Emile Schwidder and Marcel van der Linden, Leiden University Library, Nationaal Archief in the Netherlands—particularly Julinta Hutagalung, Arsip Nasional Republik Indonesia (ANRI)—especially Jajang Nurjaman and Nadia Fauziah, Perpustakaan Nasional (Perpusnas), and the British Library.

I also thank Audrey Kahin, Harry Poeze, Marieke Bloembergen, Gerry van Klinken, Herlambang P. Wiratraman, John Riddell, Fredrik Petersson, and Matthias van Rossum who have generously shared important archives and documents from their personal collection.

The following individuals provided invaluable assistance without which this book would not be in its current form. Muhammad Yusuf reproduced all of the cartoons from *Sinar Hindia/Api* newspapers from 1918 to 1926 that are displayed in this book. Adam Jessup produced the GIS maps that illustrate the geography of resistance in chapter 3. A colleague (who must remain nameless) helped collect and translate archival materials at the Russian State Archive of Socio-Political History in Moscow, Russia. Amalia Astari assisted me with Dutch archives at ANRI, and Hanum Tyagita and Herdiana Hakim shared scanned literary resources from Indonesia. Michael Needham of Humanities First served as one of my main interlocutors lending a fresh pair of eyes throughout all the drafts as I was developing this book. My sincerest thanks to all of them.

The pursuit of knowledge is a collective labor. My heartfelt gratitude to Pepijn Brandon, Bart Cammaerts, John Downing, Carla Jones, Tim Oakes, as well as Moustafa Bayoumi and our FFPP group for reading sections of this book and providing instructive comments. Even though we had not met in person, John Sidel read a draft of this book in its entirety and shared critical insights and advice that helped transform a lot of key ideas during the revision stage. I am truly appreciative of his generous support.

Pursuing an academic career can be a solitary journey. Conversations and debates with colleagues and friends made this road less bumpy and more like an adventure. In the early stages of this book, the late Om Ben Anderson inspired me to be especially attentive to taken-for-granted routes and absences. Harry Poeze provided guidance to navigate and interpret archival sources in the Netherlands during my fellowship at KITLV. At these stages too, I am grateful to have crossed paths with Indonesians who have been in exile since 1965, including Om Sardjio Mintardjo, Om Wardjo (Sarmadji), Pakde Kuslan Budiman, and Pak Ibrahim Isa—all have since passed away—and their children too, including Om Gogol and Ninik. Listening to their stories, accompanied with Pak Min's famous *sop buntut* at his Oegstgeest home, helped illuminate new understandings of the movement. In later stages, I have taken inspiration from my conversations with Max Lane, Sebastian Conrad, John Kristensen, Jan van der Putten, Sylvia Tiwon, John Roosa, Rudolf Mrázek, Meghan Forbes, and Rukardi Achmadi. The editorial board of Indoprogress.com, especially Coen Pontoh and Windu Jusuf, and the Indonesian Marxism research group, Thiti Jamkajornkeiat, Lin Hongxuan, Klaas Stutje, Xie Kankan, Paula Hendrikx, and Oliver Crawford, have over the years provided me with an intellectually challenging and politically engaging forum vital in the development of my own ideas.

I approached and pitched my book to Sarah Grossman in 2019 at the AAS Conference. I am grateful that she continued to be supportive of my book project amid the pandemic and life events that delayed my submission and revision. I must give thanks to my anonymous reviewers for their valuable and constructive insights that tremendously improved this book and to the Cornell University Press production team.

I am very fortunate to have Baruch College and the City University of New York as my intellectual home and to be surrounded by brilliant and supportive colleagues. I thank former Dean Alison Griffiths, Dean Jessica Lang, the three chairs I have worked under throughout these years—Jana O'Keefe Bazzoni, William Boddy, and Eric Gander—and colleagues at the Department of Communication Studies, especially Allison Hahn, Sarah Bishop, Stuart Davis, and Caryn Medved, for their caring and continuous encouragement.

Throughout these years, close friends and community have nurtured my soul and provided mental support as I raced to take this book to the finish line. I am particularly grateful for Margaretha Sudarsih, Alice Maher, Heather Linville, Darla Linville, Sonia Das, Sara Smith, Suzanne Weiss, William Riordan, Gino Canella, Farabi Fakih, Tri Astraatmadja, Nur Amali Ibrahim, colleagues at Prodi Inggris Universitas Indonesia, DxD & crew, @superhumans, and @asburypark for being my support system.

Lastly, my deepest gratitude goes to my family, without whom this book would not have been possible. My siblings, Rini, Farida—who also helped with the translation of German documents, and Andi, and my nieces, Noura and Hanna, are the best cheerleaders one could ask for. My late father, Agus Subijanto, was the first to open up my eyes to the cruelty of Suharto's regime and to seek an alternative understanding through furthering education. My late mother, Mei Subijanto, always pushed me to go beyond my boundaries, in her words, "*tidak ada yang tidak bisa, tidak mampu, dan tidak mungkin.*" I forever thank her for encouraging me to pursue my dreams even if that meant she had to live thousands of miles apart from her daughter. I dedicate this book to my late parents' loving memory.

This book was conceived before my daughter Daya was born; it grew as she grew. Like having another child, the book took equal amount of my energy, focus, time, and restful nights away. All these years, the book must have felt like a phantom sibling for her, continuously spoken about yet nonexistent. Now that this book is with us, please take me to your world of magic and wonder, *sayang*.

Glossary

Note on Spelling and Translations

The Malay texts when quoted are written in their original spelling. They are not transliterated into Ejaan Bahasa Indonesia yang Disempurnakan (EYD, the enhanced spelling), the current spelling system of Bahasa Indonesia. In EYD, the Malay spelling "oe" becomes "u," "j" becomes "y," "tj" becomes "c," "dj" becomes "j." For example, the words "*prijaji*," "*roekoen*," and "*ketjil*" in EYD are spelled "*priyayi*," "*rukun*," and "*kecil*," respectively.

Translations from Malay, Indonesian, and Dutch are my own unless otherwise indicated.

Frequent Malay Terms

Bergerak	To mobilize
Berkeroekoenan	To organize
Kemanoesiaan	Humanity
Keroekoenan	Organization, comradeship
Kromo	The diverse lower-class commoners without rank and status
Literatuur socialistisch	Socialist literature
Openbare	Public
Openbare vergaderingen	Public meetings
Pergerakan	Movement; the *pergerakan* era is often associated with a period of national awakening in the history of Indonesia, which the early period of communism is considered to be a part of.
Pergerakan merah	Red movement
Pers revolusioner/revolusionair	Revolutionary press
Prijaji	Member of the Javanese aristocratic and official class
Rasa	feeling, taste
Sekolah Rajat	People's School
Vergadering(en)	Meeting(s)

Woro/wiro/roro   *Woro* (Ms./Mrs.) is a title of respect before a woman's name. Sometimes, it is used before the husband's name to indicate that a woman is his wife. *Wiro* (Mr.) is used for men. *Roro* (Miss) is for unmarried women.

COMMUNICATION AGAINST CAPITAL

INTRODUCTION

The *Pergerakan Merah* as an Event of Communication

> **Our dream is to live together as one, respecting all ways of being, and overcoming all racial, linguistic, and religious difference.**
>
> —Djoeinah, "Zaman Ini" ("This Age")

On September 26, 1920, speaking in front of a thousand people in a *vergadering* (Dutch: meeting) in the Javanese mountain town of Ungaran, *woro* Djoeinah, the first female editor of the communist organ *Sinar Hindia*, described at length the suffering that workers and their families endured at the hands of *kaoem oeang* (lit. "people with money").[1] World War I and the Spanish influenza pandemic had just concluded, but the economic crisis in the Indies continued to deepen. Job losses, wage cuts, and a sustained increase in food prices created unrest among the lower-class native population.[2] In meetings and newspaper articles, Djoeinah expressed vehement criticism against both the Dutch colonial government and the capitalists that colluded in exploiting and suppressing the workers and the natives. She invited lower-class Indonesians to *bergerak* (mobilize) in a *keroekoenan* (organization). Though she spoke fiercely against colonialism and capitalism, Djoeinah did not call for violence. Instead, she advocated for the workers to fight for liberation in the name of *kemanoesiaan* (humanity). Her dream was to connect and form fellowship with others without the barriers of racial, linguistic, and religious differences. Under the Dutch colonial government, life in the Indies was highly segregated. Europeans received the highest political, social, and economic benefit, while lower-class natives received its crumbs. Between them were the Chinese, Arabs, and other Asians who were taken advantage of to serve European interests. Djoeinah challenged the traditions and moral platitudes imposed by Javanese feudal culture, religion, and Dutch colonialism that divided humanity along identity differences.

1

During the 1920s, lower-class people in the Dutch East Indies (now Indonesia) were united in a popular, archipelagic, and global revolutionary movement they called the *pergerakan merah* (red movement). This book tells a story of the processes through which ordinary people organized, led, and mobilized the *pergerakan merah*. It was an age in motion in which labor unions, political parties, and social organizations mobilized a united front under the umbrella of communism. This was a novel and exciting period in which anticolonial resistance was joined with a strong pursuit of knowledge, a desire to connect with other workers and the oppressed in other countries, and a celebration of cultures of resistance. Rather than resorting to weapons and warfare, the *pergerakan merah* developed collective, nonviolent actions around new emerging communicative technologies and practices.

The *pergerakan merah* was an event of communication. It was a period shaped by the evolution and formation of enlightenment ideas, first brought through colonialism, and the resulting contours of communication networks laid in the Indies in the century prior. By 1920s, circulating enlightenment discourses and communism merged at a time when lower-class people newly found access to modern means of communication. This period allowed women, children, and people of diverse races and ethnicities to take over enlightenment ideas, infuse them with radical and emancipatory communist rhetoric made ordinary, repurpose existing communicative technology and practices into means of struggles, and use them as weapons to campaign anticolonial resistance. This activity around communication is the movement's central strategy and is one of the consequential contributions of these marginalized figures—marginalized only in the historiography but not as agents of history, shifting techniques of anticolonial mobilization from warfare to modern forms of communication. Hence, the period of the *pergerakan merah* culminated in an inchoate but consequential movement as an event of communication.

This book moves away from leader-, party-, and formal event–centered narratives toward the everyday processes of ordinary individuals within the movement. If it was mobilized by mass followings of lower-class commoners, what did that process look like? Why do we only hear stories of the male leaders? Why do we not know about the stories of women, children, and the diversity of identities of lower-class people? More importantly, since the lower-class natives were mostly illiterate and poor, why and how did they organize in modern organizations and produce modern forms of communication? Drawing from newspapers, government reports, intelligence reports, Communist International (Comintern) files, and personal memoirs, it uncovers the democratic tradition of what I call "revolutionary communication," which consists of communication practices and technologies that are produced as anticapitalist and anticolonial mobilization.

This includes *openbare vergaderingen* (public meetings, OVs), *pers revolusioner* (revolutionary press) and print matters, Sekolah Rajat (People's Schools), communist songs, as well as sailors, railway and port workers, and postal service workers as messengers.

The diversity of lower-class commoners who mobilized the *pergerakan merah* was equally reflected in the languages they communicated in. Besides local or ethnic languages, Low Malay, which had been a lingua franca of Malay people in the archipelago, remained the dominant language of these speakers. At first it was used by people of various races for trade and commerce; then it became a marker of indigenous people. Nevertheless, Low Malay continuously grew adopting new expressions as new ideas emerged and new languages came into contact such as during the *pergerakan merah*. In the movement's revolutionary newspapers, meetings, schools, and letters, Low Malay was the main medium of communication. Depending on the audience or target readers, however, local languages—such as Javanese and Sundanese—and Dutch were often used. Code mixing was also common in which Arabic, Chinese, and Russian expressions were used in their communication; these expressions were often later absorbed into Malay lexicon. Low Malay formed the base for Bahasa Indonesia, which was formalized after the Indonesian national independence.

As the specter of communism loomed over the Dutch East Indies, popular resistance against colonialism became intertwined with a project of enlightenment. The enlightenment rhetoric that had spread through Dutch colonial infrastructures intermingled with the nascent global communist movement after the conclusion of the Russian revolutions in 1917. *Woro* Djoeinah's dream to push for universal human emancipation to remove colonialism and other old traditions reflected this broader novel spirit characterizing the period. The production of revolutionary communication in the *pergerakan merah* resulted from this global emancipatory spirit.

Beyond Borders: Revolutionary Communication as an Approach

The task of this book is to explore revolutionary communication of the *pergerakan merah* as forms of organizing and mobilizing anticolonial resistance in the Indies and to situate it in a wider-angle lens within a broader archipelagic, transoceanic, and global frame. The idea of revolutionary communication having emerged out of the experience in colonial Indonesia is rooted in existing literature on radical media and democratic communication. The communications scholar John Downing refers to radical media as "media, generally small-scale and in many different forms, that express an alternative vision to hegemonic policies, priorities, and perspectives."[3] They become alternative media because they purposefully respond to oppressive and hegemonic power and provide creative outlets for marginalized people both in helping them understand oppressive power structures and gain voice in society.[4] Chris Atton expands this understanding of alternative media "beyond simply providing a platform for radical or alternative points of view . . . to enable wider social participation in their creation, production and dissemination than is possible for mass media."[5]

Radical media exemplify the British cultural theorist Raymond Williams's concept of democratic communication. According to Williams, one needs to approach a theory of communication in terms of "how communication relates to community, how it relates to society, what kind of communication systems we now have, what they tell us about our society, and what we can see as reasonable directions for the future."[6] Democratic communication is "something that belongs to the whole society, that it is something which depends, if it is to be healthy, on maximum participation by the individuals in the society," dispersing itself into many different and independent systems owned by the public not as a way to control people or making money out of them but as true open channels of participation.[7] It is, hence, distinct from the idea of populist communication—a top-down form of communication—in which the term "people" becomes a rhetorical construct used by certain groups to appeal to a population and shape their opinion. Democratic communication is a bottom-up form of communication generated through an active participation of individuals in the society. In this context, media of communication serve as a space for this genuine, maximum participation of the individuals, and it ensures that such participation is encouraged and needed for democracy. Revolutionary communication, much like radical media, becomes an alternative channel for marginalized ordinary natives in the Indies, whose voices remained absent in dominant and mainstream media at the time, to resist and fight for their freedom and in practice it adopts the characteristics of democratic communication. However, compared to the broad term of radical media, revolutionary communication develops specifically as a set of technologies and practices that express working-class struggle and serves as an organizing mechanism to mobilize an anticapitalist and anticolonial movement.

This book approaches communication in terms of sociotechnical systems.[8] It identifies the communicative processes in creating the conditions of both constraint and possibility for the emergence, development, dynamics, and success or demise of a movement. In these processes, human agency is conditioned by the existing communicative systems to mobilize resistance. In other words, human creativity to adopt and make something new, such as communicative practices of resistance, is facilitated by the existing systems. However, they do so by assigning to these communicative systems new function, form, and operation. This process of communication in a social movement then creates what I call a "circuit of struggle" in which available systems are turned into an opportunity to create new systems with different functions. Communication as a circuit of struggle involves a "process of identifying and developing emancipatory possibilities immanent to existing conditions in order to enliven them," while surpassing the existing conditions to create a new condition.[9] Circuit of struggle resonates with Nick Dyer-Witheford's use of "circulation of struggles," which he develops from Karl Marx's concepts of "circuit of capital." He argues that "capital depends for its operations not just on exploitation in the immediate workplace, but on the continuous integration of a whole series of social sites and activities . . . [which] may also become scenes of subversion and insurgency."[10] My use of circuit of struggle is aimed to probe and highlight in detail the processes involved in turning the circuit of capital into that of struggle. This idea of circuit of struggle helps us understand how systems of resistance often develop intimately along with, if not out of, those of exploitation through creative processes of modification, adaptation, and repurposing such as the case of revolutionary communication produced by the *pergerakan merah*.

Communication history is a story in which human creativity is integral to the making of communications practices and technologies both as a response and a consequence. Humans create and transform their communication technologies as a response to a particular condition of their lifetime. These technologies therefore inhere human creativity, needs, intention, and value, while also shaping behaviors, actions, and ways of thinking. They limit humans in that they pose as constraints but also become conditions of possibility for a new system. The new emergent system, however, might result from the intended and unintended consequences of this creative process. For example, Dutch colonizers did not intend for newspapers, shipping lines, and Western education to provide the lower-class natives with a language and means for articulating their own values, ideals, and hopes of liberation. Likewise, the revolutionaries did not intend to spark repressive actions by the Dutch colonial state when they centered communication as their key means of mobilization.

Approaching the revolutionary communication as sociotechnical systems allows us to see the *pergerakan merah* in its mundane, ordinary day-to-day processes of its production. This book especially pays attention to materials that often are considered secondary, unimportant, and partial.[11] Despite often being considered biased sources, newspapers, for example, chronicled history in the making, capturing events as they unfolded, often with uncertain outcomes. This inherent focus on the present rendered their content transient and riddled with contradictions. Rather than a flaw, this quality is crucial in grasping the *pergerakan merah* as an ongoing "process," offering nuanced perspectives absent in post-event official reports. Songs, newspapers, letters, and memoirs are sites where *rasa* (feeling) of the past could be shared and felt in an intimate way, and express the reasons why these revolutionaries struggled for their emancipation. The lived experiences, sufferings, hopes, and dreams could be vividly heard in Djamaloedin Tamin's handwritten memoir, *woro* Ati's court defense, and an unknown communist lady who sang Mariana in a train en route to a *vergadering* (meeting).[12] This is how ordinary communists within the *pergerakan merah* was lived and experienced. This is not a narrative of elites and leaders versus commoners and followers; rather, it situates the historical agents—including leaders—in their human conditions. *Rasa* is a way to embrace the mass lower-class members of the movement. Even though we cannot register each of their names or imagine how they looked, we can still delve into their suffering, hopes, and dreams. It lends color to a faceless revolution.

This book also approaches communication as sociotechnical systems in terms of its connection between transport and social networks as networks of mobility and sociability. The ideas of mobilization and socialization are at the core of social movement theories, but both necessitate communicative processes. The communicative modes of mobility, such as the spread of ideas and propaganda, the mobilization of people, and the movement of activists to connect and network across different places, manifest struggles to occupy different places and spaces. Similarly, the communicative modes of sociability, such as the creation of collective identity, of collective goals and actions, and of collective organization through the use of symbols, identity markers, framing, and discourses, exhibits the creation of both mental and physical spaces of contention.[13] In this case, the idea of communicative sociotechnical systems helps bring together disparate aspects of social movements. We must move beyond thinking of the role of media in terms of tactics and strategies and consider them instead as integral parts of the making of a movement in terms of networks of mobility and sociability.[14]

Thinking of communication in terms of the connection of transport and social networks necessitates us to think of the movement in a global framework. The media scholar Nick Couldry describes communication and media as "infrastructures of connection" to inquire into the distinctive and universal ways about the world that they help create.[15] Communication practices and technologies integrate the world and facilitate mobility and sociability of people, ideas, and things. In terms of the global communist movements, as John Sidel convincingly explores, market, commerce, and communication networks facilitate the arch of cosmopolitanism for the revolutionaries in Southeast Asia in this period.[16] This book includes an analysis of the global, as it is instantiated at the local experiences. There is essentially no contradiction between local and global, as Sebastian Conrad suggests. In fact, the local and the global emerge in relation to each other. It is only through the instantiation and manifestation of the global in local practices that the global is produced and reproduced.[17] Communism became both local and global as it came into contact with Islam, Javanese identity, and other values that mattered for the participants in colonial Indonesia. The global making of communism at a local level is active and creative. However, the production of revolutionary communication within the framework of enlightenment illustrates more vividly the interplay of the global and the local. This book, therefore, moves away from earlier scholarship on the period by highlighting this global–local entanglement.

An Age in Motion Revisited

The *pergerakan merah* was strikingly strong, early, powerful, and important in the history of communism in Asia and more generally in terms of anticolonial mobilization across the world. Partai Komunis Indonesia (the Communist Party of Indonesia, PKI) was the first communist movement to be founded in Asia beyond the borders of the former Russian Empire. It was the oldest major Indonesian party and the first to carry the name "Indonesia" (previously "Hindia")—a name that reflected the nascent national identity of the natives.[18] It was also the earliest, largest, and most organized anticapitalist movement in the history of Indonesia and Southeast Asia. The movement solidified the transition of anticolonial resistance in the Indies, from traditional, sporadic, and local to systematic, coordinated, and widespread; its networks spanned across the archipelago and beyond the Indies.

However, the *pergerakan merah* was not just significant because it changed the forms of anticolonial mobilization from its traditional past. It also paved the way for the National Revolution that occurred two decades later. The *pergerakan merah* was able to both mobilize a mass of lower-class commoners and to radicalize the popular struggle for self-rule and independence. Previous modern organizations of the National Awakening era, such as Boedi Oetomo and Indische Partij, campaigned for natives' welfare and progress, but their leadership and membership continued to be limited among a few native elites. Sarekat Islam (Islamic Union, SI), however, was able to attract and connect peasants, workers, and mass followings in rural and urban areas in Java and other islands. When SI was penetrated by the leadership of the Indische Sociaal-Democratische Vereeniging (Indies Social-Democratic Association, ISDV), which later became the PKI, their followers turned their campaigns to broader issues of social justice,

emancipation, human rights, and the struggles to oust colonialism and capitalism from the Indies.

Even when we observe the movement from the center of the Dutch colonial government in Buitenzorg, West Java, it is clear that PKI and the *pergerakan merah*'s anticolonial mobilization posed new, direct, and real threats to the
colonial government.[19] In response, the Dutch state formed new institutions to
protect its colonial interests from the red scare, including the Political Intelligence Service, police institutions, laws, and regulations. A coordinated effort in
policing and suppressing the global red scare also led to cooperation between
Dutch, British, and French colonial empires in the region. When the *pergerakan
merah* was banned following the revolts of 1926 to 1927, thousands of the *pergerakan merah*'s followers were exiled to a newly established concentration camp in
Digoel, an isolated area in a malaria-ridden Papua island about 4,000 kilometers
from Java. The movement darkened colonial policy and led to an oppressive and
repressive infrastructure of policing.

This red scare intimidated the government due to both the movement's radical character and its vast reach. Audrey Kahin and Steve Farram expand the Java-
centric account of the movement in earlier literature and unearth the dynamic
of the *pergerakan merah* in Sumatra and the Timor islands, showing how the
movement expanded well beyond Java. Beyond the archipelago, PKI's network
and activities spread throughout Asia and Europe in sea and land alike. Ruth
T. McVey's *The Rise of Indonesian Communism* shows how the PKI's strategies
and tactics were important both within the Indies and internationally throughout Comintern networks. Comintern, based in Soviet Russia, played a role in
supervising communist movements around the world, including the recognition of the establishment of a communist party in each country, the approval of
its committee, providing funds, delegation of agents, and receiving and training of communists from other countries. Harry Poeze's work on the Indonesian
revolutionary leader Tan Malaka demonstrates his important roles in shaping
Comintern's policy on anticolonialism and on Islam and the Muslim world. In
fact, communism did not become relevant in the colonies until it aligned with,
and took serious accounts of, anticolonial resistance. In other words, communism became universalized through its adoption of anticolonialism. Along with
other Indonesian communist leaders, such as Semaoen and Darsono, Tan Malaka
expanded the reach of Comintern networks in Europe, China, Southeast Asia,
and elsewhere both in his clandestine activities, as well as in writings.[20] However,
how exactly this international side of PKI was entangled with the day-to-day
processes of mobilization by members who never left the Indies remains a puzzle.

Takashi Shiraishi's *An Age in Motion: Popular Radicalism in Java, 1912–1926*
opens up new ways to understand the movement. He demonstrates that the

pergerakan merah was, in fact, organized beyond party lines and mobilized by a network of political parties, social organizations, and labor unions. Additionally, he argues that people in the *pergerakan merah* expressed the new consciousness not just through formal forms of organizations but also through modern strategies and languages in newspapers, rallies, strikes, unions, and ideologies. While Shiraishi successfully identifies important rank-and-file local leaders, such as Hadji Misbach, Tjokroaminoto, and Marco Kartodikromo, ordinary mass followers remain anonymous.[21] Moreover, his inland and local focus on Surakarta, Central Java, leaves many unanswered questions about the international aspects of the movement.

The study of Indonesian communism, Marxism, and socialism from a global lens has grown in recent years, and the dominant national and nationalist framework of earlier scholars is being complemented with a global view of the movement. John Sidel and Tim Harper's recent publications unearth the cosmopolitan and global connections and infrastructures of integration that set the stage for early twentieth century Southeast Asian revolutionaries.[22] For the first time, we are learning how the global overground and underground network of mobilization of the Indonesian communists was linked to movements in neighboring Southeast Asia and distant lands from Baku and Istanbul, to Shanghai and Tokyo. Recent research on the fugitive policing networks across Southeast Asia, intellectual exchanges across Asia, Europe, and the Middle East, Indonesian students' activities in the Dutch metropole, and translation and transliteration likewise demonstrate the cosmopolitan and global aspects of the movement.[23] Over the last few years, McVey's 1965 inquiry about whether Indonesian communism was local or foreign in origin has become irrelevant. The world of ordinary Indonesians during the *pergerakan merah* was already global and in constant change, although the majority of them were poor and illiterate and did not have the means to travel abroad. They did not live in isolation from the world; they actively participated in the making of the global order.

In the 1920s, Indonesian communists did not imagine the goal of their liberation within national boundaries; rather, they envisioned themselves as part of an international community and solidarity of workers, peasants, and the oppressed. This movement was highly global, cosmopolitan, and media savvy, and led by lower-class ordinary members. These people took a leading role in the development of the movement through the collective production of revolutionary communication, especially in times when the main leaders were already in exile. As a part of a chain of global commodity production and distribution, Indonesian communists created a place for themselves in a global society through the language and rhetoric of enlightenment and communism, challenging local and traditional authority—religious, patriarchal, and royal—as well as colonial power.

Being communists did not entail betraying local cultures or making themselves inauthentic; rather, they were keeping pace with the global society.

Red (e)nlightenment: An Event in the History of Mediation

Woro Djoeinah saw emancipation as coming not from the transcendental—God or mystical spirits—but rather from something that was immanent: believing in the human capacity to both understand the world and to change it. This reflected the broader structure of feeling of the period in which ordinary communists expressed the new age and consciousness in writing and speeches. In criticizing and analyzing the conditions of their suffering, they sought what Nick Nesbitt calls "*universal* emancipation (as opposed to, say the emancipation of white, male, adult property owners)."[24] Lower-class native Indonesians wanted to overcome unequal and unjust conditions that were rooted in colonialism, including patriarchy, religion, and aristocracy, but they did so by appealing to human universal and unconditional rights to coexist regardless, or perhaps alongside, their differences.

The movement's emancipatory spirit expressed a combination of aspirations from the Enlightenment era and the novel communist forces. Tracing the nature of this spirit situates revolutionary communication within a broader historical context. This book explores the dialectic of enlightenment as it was manifested in the production and practices of revolutionary communication. It explains how the *pergerakan merah* in the 1920s borrowed, modified, and adopted enlightenment rhetoric, and combined it with communist rhetoric as part of their emancipatory spirit, creating a red enlightenment. Red enlightenment is a unique event of mediation that shapes techniques and strategies of mobilization during the first popular and radical anticolonial movement in colonial Indonesia. Notice that enlightenment here does not refer to the Enlightenment era (with a capital E). Rather, enlightenment means discourses, values, practices, and rhetoric of enlightenment that continues to live and to evolve beyond eighteenth-century Europe. Therefore, when referring to the Enlightenment era in Europe in the eighteenth century and the ideas that directly emerged out of it, this book uses "Enlightenment," but, beyond this period and locale, it is lowercase. What is equally important, however, is to understand enlightenment within the framework of communication.

Existing scholarship has inquired into the Enlightenment and the history of communication. Clifford Siskin and William Warner, in *This Is Enlightenment*, explain that "Enlightenment is an event in the history of mediation." Knowledge necessitates mediation. The Enlightenment occurred and was manifested as an "event" in myriad communicative forms, including newspapers, magazines, pamphlets, public lectures, salons, and coffee houses.[25] Lawrence Klein additionally argues that the Enlightenment is not defined by "a set of doctrines but by a set of communicative practices," conversation, politeness, and sociability that were conducted using conversational ideal of equality, reciprocity, and the ease of informality.[26]

Radical Enlightenment ideas animated democratic revolutions, and the democratic emancipatory characters are most vividly demonstrated in the democratic communication that these revolutions produced. In the French Revolution, the Paris Commune, the American Revolution, the Haitian Revolution, and the Russian Revolution, Enlightenment ideas gave birth to new ways of organizing and mobilizing by means of assembly, association, literature, and literacy.[27] As Michael Warner wrote regarding the age of the American revolution: "The eighteenth century was remarkable for its literature and revolutions."[28] Likewise, the seventeenth-century civil war in England "unleashed an explosion of popular energy, which found expression above all in a phenomenon relatively new to British political culture—tens of thousands of pamphlets, petitions, magazines, leaflets, and handbills."[29]

In spite of their differences, the working classes in the plebeian public sphere, French salons, English pubs, Haitian clandestine social networks of mulattos and sailors, and the Black press all struggled for broad but palpable goals: freedom and liberty. The historical development of democratic communication is directly tied to the global project of enlightenment as a social movement. This is especially true since the institutions of the Enlightenment public sphere produced networks of communication and sociability, enabling people to overcome locally embedded identities and to imagine themselves as part of a larger political community. As Melton suggests, "[i]n the end the institutions of the Enlightenment public sphere were the product of an information revolution, and information revolutions."[30] Universal emancipation is thus expressed and manifested in communication practices.

In the context of Dutch history, Enlightenment connects in an intimate way with the development of certain contour of communication networks. Joost Kloek and Wijnand Mijnhardt, in *1800: Blueprints for a National Community*, describe the revolutionary decades in the 1800s that shaped and developed the Dutch Republic and its civil society.[31] It illustrates how the French Revolution and the Enlightenment in Europe and the Americas helped define conversation on human emancipation and the Rights of Man in the Netherlands. Kloek and Mijnhardt argue that the Enlightenment in the Netherlands gave birth to, and in turn developed, through the contours of a communication network. This network included free debate, the printed word and nationwide press, sociability (clubs, salons, and coffee houses), the culture of the educated classes, and the birth of public opinion. Roughly a century later, during the *pergerakan merah*, similar contours of a communication network were assigned to revolutionary communication.

Kloek and Mijnhardt explain how the Enlightenment and the eighteenth-century Dutch culture shaped the Dutch republic and its civil society, but they do not mention Dutch colonialism and its empire. René Koekkoek, Anne-Isabelle Richard, and Arthur Weststeijn observe that, while the constitutional, economic, institutional, and legal design of the Dutch empire was created within the context of Enlightenment cultures of knowledge and ideas, literature on Dutch Enlightenment and its political culture and thought more broadly has not been put in conversation with those on the Dutch imperial world. This sharp criticism warrants attention, especially given that the colonies are often described from a local and national framework that detaches the colonial periphery from the Dutch metropole. The authors suggest a much closer connection between the making of Enlightenment Dutch culture in the metropole and its oppressive and exploitative colonial expansion, and the questions they pose deserve to be presented here in their entirety:

> As the colonial system of trading companies came to an end and colonial governance was transferred to the state, what moral and political principles were invoked to justify or criticize colonial rule and exploitation in this period? How were policies regarding non-western peoples recast in light of Enlightenment theories of historical progress and civilization? What was the impact of eighteenth-century liberal economic thought concerning trade and labour on the political-economic design of the empire? In addition, how were Enlightenment ideas and concepts applied, appropriated, enriched, tested, amended or refuted once they transferred beyond their European and Dutch origins? And finally, how did colonial subjects and local populations respond and adapt to, as well as resist these innovations in imperial political thought, practices, and culture?[32]

These questions require a broader approach that connects local instantiations of ideas and practices in the Netherlands and the Indies with each other. Even though the approach is beyond the scope of this book, the book provides case studies from the Indies that shed light on this connection.

Recent publications on the global history of the *pergerakan merah* in the Indies generate ways to connect communist anticolonial resistance with enlightenment history. Tim Harper's *Underground Asia* discusses the "fury of enlightenment" in which European civilized standards descended from the Enlightenment era were challenged by a growing number of young Asian intellectuals traveling and acquiring education in Europe. These scholars used these standards against the masters to question their colonial practice and empire. During his study in Haarlem, the Indonesian revolutionary Tan Malaka found Thomas Carlyle's *The French Revolution* "a resting place for my weary, questing thoughts" taking inspirations from French Revolution for his anticolonial ideas and politics.[33] Likewise, engagement of Enlightenment ideals from other Southeast Asian revolutionaries—from Jose Rizal to Ho Chi Minh—is well documented.[34] The colonialists and anticolonialists shared the same intimate paths toward different outcomes. The manifestation of enlightenment in contours of a communication network is vividly uncovered by Rudolf Mrázek's *The Complete Lives of Camp People*. In a section titled "Light," he describes how concentration camps were spaces of enlightenment. In the aftermath of the communist revolts of 1926 to 1927, communists exiled in Digoel brought books and Western literature to the camp, continued to read and write, created library and schools for the children, and organized *toneel* (modern theater) performances, just like they did when the movement was alive and thriving.[35]

Chapters 1 and 2 of this book begin an exploration of revolutionary communication by delving into the enlightenment ideas that were alive in the Indies at the turn of the nineteenth century through the beginning of the twentieth century when the *pergerakan merah* emerged. Chapter 1 explores a wide-angle view of the communicative infrastructure and connection of integration that creates a backdrop to the global lived experiences of ordinary Indonesians. It explains how enlightenment as sources of colonialism drove the development of communication networks, and how the development of modern communication network led to the changes of the class structure of the native population who would later lead the movement in 1920s. This provides a context for chapter 2, which delves into how and why the new class of the educated *kromo* (the diverse lower-class commoners without rank and status) took upon enlightenment and combined it with global communist rhetoric to mobilize anticolonial struggles.

The subsequent chapters examine the manifestation of enlightenment ideas in the production and practice of each form of revolutionary communication. Chapter 3 provides rich and original visual and statistical evidence for the expansive spread of the movement, the diversity of its participants, and new important timelines prior to the revolts of 1926 to 1927. Analyzing a set of historical geographic information system maps based on 865 reports of communist OVs that were held across the Indies archipelago from 1920 to 1925, these maps illustrate that the movement was mobilized, for the first time, across widespread geographical areas in the Indies archipelago and that it was organized across different cultural borders and identity markers. This chapter provides new evidence that the 1923 jailing and exiling of its main leaders in fact led to the peak years for the movement, in both radicalization of the movement and in the anticommunist reactions.

Chapters 4 through 7 discuss OVs, People's School, *pers revolusioner*, and sailors as different sources of revolutionary communication. Chapter 4 delves into the communist OVs and schools to understand the making of politics, here understood as "the common cultures of resistance," in its ordinary setting by ordinary people in their everyday lives. I trace the global in the local, the mundane in the revolutionary, and the ordinary in the extraordinary processes produced by the movement by explaining the three social functions of OVs: namely as entertainment, as educational institutions, and as cultures of defiance. *Vergaderingen* are presented as a site where uneducated peasants, workers, and women learned to debate and make collective decisions and to use the vocabulary of universal emancipation in their daily exchanges. Here the creation of a community organized around equal and collective cultures replaced those previously organized around religion and monarchy.

Chapter 5 recovers the tradition of *pers revolusioner* and situates it within the history of anticolonial struggles in the Indies by examining the production and development of the revolutionary newspaper *Sinar Hindia*. It reveals how *Sinar Hindia* not only embodied the anticolonial struggle but also became a voice for a project of enlightenment in the colony, including the importance of journalism, science, and reason. By uncovering the revolutionary paper's own discourses of enlightenment and revolutionary struggle, the chapter sheds light on the role of the press in the production of enlightenment ideas and practices in colonial Indonesia.

Chapter 6 provides a logical chronology for when OVs, communist schools, and *pers revolusioner* ended and when the clandestine movement of sailors began. The chapter discusses anticommunist propaganda, actions, and regulations around revolutionary communication. Using the concept of communication as a "circuit of struggle," I argue that, during the *pergerakan merah*, the colonial state sought to pacify the anticolonial resistance for the first time by releasing repressive laws and policies around three forms of revolutionary communication discussed earlier—OVs, People's School, and *pers revolusioner*. In the name of "public peace and order," this evolution of the legal measures on communicative practices and technologies reflects the broader shift of the Dutch colonial state from an ethical state to a police state. This shift shaped a legacy of media law and policy through postindependence Indonesia wherein public communication focuses on order and discipline. This chapter shows how, in lieu of warfare, modern communication and law—products of enlightenment—becomes a circuit of struggle between the colonial state and the colonized.

Chapter 7 delves into the production of enlightenment communication as "the other labor," activism work that occupies and repurposes formal work, time, and space. It provides evidence that the Indonesian communist movement did not completely end after its ban in 1927. It narrates the clandestine activity of communist sailors overseas following the 1926 to 1927 revolts. Within the framework of labor, it argues that revolutionary works and international solidarity were developed through mundane, daily, frequently boring, and, at times, dangerous work, and rarely as grandiose or celebratory endeavors. This chapter challenges the notions that the early communist movement ended after the revolts of 1926 to 1927 and that the traditional geographical scope of the movement was limited within the borders of the Indies. Through the activities of communist sailors, the movement was organized beyond the archipelago and, after the aforementioned revolts, the evolving movement continued clandestinely until PKI was revived again during the National Revolution of 1945 to 1949.

I close the book with reflections on the legacy of the *pergerakan merah* for anticolonial and anticapitalist struggles. This book recovers an important story of the place of revolutionary communication under the *pergerakan merah*. It reveals that its role was central in both the birth of national liberation against Dutch colonialism and in the global history of enlightenment. This history helps us consider the importance of democratic communication as part of a wider resistance movement for equality and justice historically in Indonesia, and underscores the importance of global solidarity in the fight for universal emancipation and human rights today.

MEANS OF COLONIALISM AS A MEANS OF RESISTANCE

The postal connection, the connection with our Javanese sailors, the political one and others are there of big importance for the revolutionary movement in Indonesia.

—Semaoen, Moscow, November 15, 1924

Innovation in transport technologies played a significant role in the success of Dutch colonialism. Within 200 years of its establishment in 1602, the Dutch East India Company (Vereenigde Oost-Indische Compagnie, VOC) had a fleet of more than 1,500 ships, making it the largest mercantile shipping company in the world. With its advanced shipping and navigating technologies, the company dominated Asian trade and became one of the landmarks of the Golden Age— the period of economic prosperity in which the Netherlands conquered and led the world's markets.[1] Though the VOC ceased to exist on December 31, 1799, the Dutch continued to take a leading role in innovation in maritime technologies. By the turn of the century, contours of a communication network in the Dutch Republic were shaped by the novel waves of Enlightenment spirits that took over Europe. This is especially noticeable in the emergence of free speech, free press, and the culture of educated and cosmopolitan classes that were facilitated by the advancement of transport technologies.[2] The communications scholar Armand Mattelart demonstrates that modern communication networks began early in the Enlightenment era and that their global nature was spawned by Enlightenment ideals of universalism and liberalism. Dwayne Winseck and Robert Park, however, add that the international reach of modern communicative infrastructures was accompanied by the strengthening of colonial empires and cultural domination.[3]

The Dutch brought Enlightenment ideas to the colony to extend their colonial power, and they continued to be practiced in the Indies in the 1800s through the beginning of the 1900s. These philosophies were most vividly evident in the development and evolution of modern communication networks. The invention of steamships allowed the implementation of regular shipping lines between the Netherlands and other regions in Asia and the development of railway lines between the hinterlands and port cities in the Indies, enhancing Dutch economic and political power. This story of transport technologies and networks as the main pillars of the expansion of Dutch power in the East Indies archipelago aligns with a quote that is attributed to Napoleon Bonaparte: "The larger the empire, the more attention has to be paid to the major means of communication."[4]

This chapter accomplishes two feats. First, it explores how enlightenment as a source of colonialism motivated the development of modern communication networks. This development shaped new modes of sociability and mobility of ordinary Indonesians that was both intra-archipelagic and global, and it also led to the changes of class structure of the native population who became the main mobilizers of the first popular and radical anticolonial movement in the Indies in 1920s. Within this movement, communication workers—sailors, postmen, dockworkers, and railway workers, among others—played leadership roles because their specific conditions as one of the most educated and mobile groups enabled them to adopt and repurpose aspirations of modernity and enlightenment. This explains why the new class of the educated *kromo* took upon enlightenment ideas and global communist rhetoric and turned them into a red enlightenment to mobilize anticolonial struggles that were national, archipelagic and global (this concept is explored in chapter 2).

The chapter's second task is to shed light on the processes that allowed the means of colonialism to become the means of resistance. While transport networks were vital for the creation of the Dutch colonial state, its economic productivity, and its political sovereignty, little attention has been paid to their eventual roles in shaping and transforming the mobility of ordinary people in the colony.[5] This chapter shows, for the first time, how the transport networks supported the economic and military success of the colonial state by connecting expansive areas in the Indies while also changing the scope and direction of ordinary peoples' mobility. This new mobility explains the changes in the class structure of the natives and the ways that anticolonial movements were organized and mobilized.

Recognizing the often Janus-faced nature of technology and power, the chapter also demonstrates the ways in which technology that was designed to strengthen and improve a colonial empire became one of the sources of its own defeat.[6] The development of transport networks, intended to strengthen Dutch control and extend capitalist infrastructure over this vast colony, became the conditions of possibility for ordinary people to organize and propel the *pergerakan merah* (red movement) against colonialism across widespread geography and identity markers. By tracing the modernization of the Dutch colonial state in the areas of modern communications, including transport and social networks, this chapter reveals how colonial communicative infrastructures became connected to anticolonial resistance in intimate and direct ways through ordinary workers who helped create them.

Multiple Enlightenments for One Colony

The Enlightenment era and its theories of progress and civilization shaped Dutch policies and outlook regarding its colonized subjects, and the liberal economic thought of the eighteenth century influenced the subsequent political economic design of the colony in the long 1800s. In the words of the Dutch historian W. F. Wertheim, "[though] the call for liberty, equality and fraternity echoed but faintly in the Indies, the political structure of the Dutch possessions there was strongly influenced by developments in Europe."[7] Remco Raben similarly argues that after the demise of VOC, "the Dutch empire acquired two important new elements: Christianization . . . and Enlightenment values."[8] However, no consensus has been built regarding what enlightenment meant and how it was to be implemented. A quick overview of the evolution of the ideas of enlightenment shows that there are several, at times opposing, interpretations of enlightenment among Dutch officials.

The first interpretation argues for *the Rights of Man*. Willem van Hogendorp wrote one of the earliest inquiries regarding the Enlightenment and Dutch colony. During his time in the Indies from 1774 to 1784, van Hogendorp witnessed that "'Batavian Enlightenment' first put into print its new ideas on welfare, progress, and the cultivation of the arts and sciences." A member of an old aristocratic family from the province of North Holland and a graduate of law from Leiden University, he was appointed as resident of Rembang in 1774 and remained there until 1777, when he became the second administrator of VOC's establishment on Onrust island, a major shipyard of the company. While in the Indies, van Hogendorp played an important role in advancing education, arts, and sciences, cofounding the Batavian Society for the Arts and Sciences in 1778, leading its work on promoting "agriculture and trade, the general welfare of the colony, and the channeling of its production to the benefit of the home country, as well as the arts and sciences."[9]

In his play *Kraspoekol*, which was translated and published in Batavia in 1779, van Hogendorp discusses the slavery question in Dutch possessions. He asserts that slaves in the Indies were as a "rule well treated *except for* abuses, some of them extremely barbarous, inflicted by female owners." These abuses included sexual exploitation, mostly of the *nyai* (Indonesian de facto wife) by the Dutch heads of household. According to Ann Kumar, "more than half of the population of Batavia at this time were slaves, and perhaps two-thirds of them were owned by Europeans." (The rest were owned by Chinese and Arab residents). *Kraspoekol* is generally liberal in its commentaries on moral correctness, and van Hogendorp's Enlightenment ethics are on full display throughout.[10]

In 1800, *Kraspoekol* was revised by his eldest son Dirk van Hogendorp. Dirk had been appointed as the senior merchant and commandant of Java's northeast coast in 1794, after serving as a resident of Jepara. A year later, in 1795, the Batavian Republic was established under the protection of Revolutionary France to end the company's charter and to take its assets, and the authority of the Dutch colonies was ceded over to the British. The most powerful members of the VOC resisted the revolutionary ride, but Dirk van Hogendorp chose political alignment with France and its egalitarian ideas. After the abolition of the VOC, he was appointed to the committee to decide on the shape of the Indies government. Demonstrating a liberal program deriving from Adam Smith, just like his father, his publications advocate for the separation of government from commercial interests, personal ownership of land, freedom of person, the abolition of compulsory labor, and freedom of trade.[11]

On the issue of slavery, he took a far more radical position than his father. He wrote a highly contested document proposing the emancipation of slavery before those views became popular in Dutch public opinion. The more liberal "Patriots" believed that slavery was incompatible with the Rights of Man; however, the rest of the majority argued that antislavery measures would have grave economic and political consequences in the West Indies—which was powered by slavery—and that the profits from the slave trade would be lost to other powers. As Kumar argues, when "forced to choose between ideals and material gains," the Dutch chose the latter.[12]

The majority of Dutch officials argued for an interpretation of enlightenment as *the rights of the white man*. Jan Breman explains that very few officials at the time would find it acceptable to apply the Enlightenment ideals that underpinned the American War of Independence and the French Revolution to the colonial rule in the tropics. The majority of these officials believed that the principles of liberty, equality, and fraternity are exceptions rather than norms to be upheld universally, and applicable only to the white tribes of the Atlantic community.[13] Dirk van Hogendorp was among the very few urging for this civilizing mission to be recognized, especially in the Dutch colonies. This means: "to protect these same Inlanders from all violence and hostility from without and within; to provide and assure them of civic liberty, the right to own property, protection from all oppression and the exercise of impartial and fair justice."[14] Dirk van Hogendorp's voice, however, remained part of the minority, and the interest of overseas economic profits took center stage in colonial policy. The double standards of the European powers can be seen in the subsequent colonial policy throughout 1800s. Dutch thinkers and civil society—along with their European counterparts—developed theories of Enlightenment, while their countries built empires that oppressed and exploited native subjects.

The subsequent generations of colonial officials treated enlightenment as a prime source of colonialism—a means to expand a Western style of administration and liberal economy and to apply modern infrastructure, science, and technology. Their understanding of enlightenment perceives that *the colonized and non-Western subjects could be emancipated but only once they received the necessary knowledge and institutions.* This is especially evident during the interregnum eras and the subsequent era of *Cultuurstelsel* (forced labor) or Cultivation System. Dutch officials saw colonialism as a training field for natives for progress and welfare, for enlightenment. To achieve this, modern infrastructures needed to be built, and changes in power relations had to happen. In other words, the natives needed to be closer to the enlightened Western officials than the backward native aristocrats.

In the interregnum periods, the double standard of Enlightenment was put firmly in place. During the short period of French interregnum (1806–11), "the spirit of the new age" in Europe was brought by Herman Willem Daendels.[15] He was appointed by the Dutch king Louis Napoleon—the brother of the French emperor Napoleon, who at the time ruled Holland—as governor-general of the Dutch East Indies. Daendels was known to be the first governor-general who brought European Enlightenment to the Indies in a form of an explicit recognition of native law issuing a ban on cruel and inhuman native punishments.[16] He was the first governor-general "convinced of the necessity of improving the infrastructure," including "the construction and maintenance of roads and bridges, offices, warehouses and bungalows."[17] He demanded that peasants performed unpaid services for these infrastructural works. He constructed the Grote Postweg, the main road across Java, to reduce travel time for mail and officials.[18] During the British interregnum (1811–16), which was led by the English statesman Thomas Stamford Raffles, Daendels's policy was continued to "impose European authority more firmly and thoroughly than ever before." His hope was to "destroy the power of the regents and to bring the peasantry into direct contact with an enlightened Western government."[19]

This authoritarian stance continued when the Dutch East Indies was handed over to the Dutch state again. Governor-General Johannes van den Bosch believed that "to apply to an ignorant and idle people the liberal institutions of an enlightened age is as impossible as to introduce religious toleration among blind fanatics. First one must try to enlighten their understanding, and then to improve their institutions."[20] Here Van den Bosch insinuates that the Javanese character and ways of thinking must change before capitalism can be effective.

Daendels, Raffles, and van den Bosch believed that natives could be enlightened if they learned from the master and adopted modern institutions and infrastructures. Unlike Dirk van Hogendorp's colleagues, who believed that Enlightenment was not applicable to non-Western people, Daendels, Raffles, and van den Bosch saw Enlightenment ideals as operational aspects of colonialism. The policies of the three leaders each represented the process of deepening penetration of Western government and Western economy into the Indonesian economy. The Javanese *desa* (village community) experienced the consequences of contact with the West in forms of land tax system, forced labor of natives to cultivate crops for the world market, and the extension of the Cultivation System to sawah (rice field) to alternate between sugar and rice cultivation.[21]

Elisabeth Moon argues that, even though he argued in favor of enlightening colonized liberal institutions, Governor-General Van den Bosch himself would make no efforts to facilitate or encourage this change; he was more interested in developing infrastructure and trade. To escape the financial crisis caused by the Java War of 1825 to 1830, Van den Bosch installed the Cultivation System, which lasted from 1830 to 1870. Peasants were required to use a fifth of their land to grow specified commodities for export, among them coffee, cane sugar, tea, peppers, tobacco, and quinine trees. As the Dutch government official Edward Douwes Dekker later wrote, this period of colonial suppression brought severe economic hardship for the Javanese. The Indonesian author Pramoedya Ananta Toer later observes that by the end of the Cultivation System, the Netherlands and its colony escaped the economic crisis while building railway systems in the Indies. The genocide, as Toer calls it, led to the creation of basic infrastructures for further construction of colonial communication systems.[22] C. Fasseur agrees, writing that it is the Cultivation System that first brought Java into "the world trade system, where Dutch trade and shipping formed indispensable links."[23]

The Cultivation System came under strong criticism by the 1840s and increasingly so in the 1850s, when another interpretation of enlightenment emerged: *The natives must be emancipated from colonialism*. Dekker sparked a sensation in 1859 with his book *Max Havelaar*. He criticized the Cultivation System for exacerbating the exploitation and oppression of peasants by indigenous leaders, and for the famines caused by a policy that minimized food production in favor of growing more profitable exports. Douwes Dekker's idea of enlightenment deviated from the majority of Dutch officials. Like Willem and Dirk van Hogendorp, he called for the emancipation of the colonized subjects from the abuses committed by Dutch colonialism (Dekker's ideas are discussed further in the next chapter).

After decades of exploitation and oppression of Javanese peasants under the Cultivation System, the Indies entered an agrarian reform or the "Liberal Period" (1870–1900) that witnessed deep transformation toward modernization of its infrastructures. The economic side of the Liberal Period began in the 1860s and expanded into the Ethical Policy of the 1900s. The developments in the Indies during this time were affected largely by the transformation of Western countries into industrial states.[24] During the Cultivation System, liberal critics became even more ruthless and demanded the Indies to be reopened for private exploitation. Gradually, the Cultivation System was transformed to a freer enterprise based on export production. Enlightenment spread in its liberal form, giving colonialism a humanist bent with a focus on natives' prosperity, welfare, and progress. The advancement of modern communication infrastructures exploded as a part of a total transformation in politics, culture, and economy. This transformation was crucial to explain the changes in ordinary lives of the natives, and it shaped the conditions for the emergence and success of the *pergerakan merah* in 1920s.

The Modernization of the Dutch Colonial State

The late 1800s was a period of transformation for the Dutch East Indies. New modes of governmentality, the change from the Cultivation System to a new style of colonialism, the expansion of production, the expansion of territorial control, and the massive development of transport networks in the archipelago created an integrated whole that transformed the economy, politics, culture, and social life of the colony. These types of innovations are not free of value. They are built out of specific motivations, needs, goals, and interests of social groups and individuals, and once they are materialized, they shape and construct social spaces "from the global to the everyday."[25] To think about transport as technology of mobility in relation to colonialism reveals the spatial production of colonial power and its ensuing implications for ordinary lives within the colony.[26] The history of the shipping lines and railways, two of the most important transport networks in the Dutch East Indies, explains how colonial interests and needs were embedded in these machineries.

For about thirty years, the interinsular steamship connections between the many scattered islands were served by British lines because the Dutch colonial state lacked modern infrastructure and technologies. In fact, until 1886, the Dutch did not have a direct steamer connecting Sumatra and Java islands. Tobacco freights from Deli in Sumatra, for example, had to stop in British Singapore before continuing to Batavia. The freight rates to Singapore were much lower and the facilities for ships' maintenance were more advanced. The access from Sumatra to mainland Southeast Asia before and during the colonial era explains Sumatra's links through trade, as well as the migration of workers to Singapore and the Malay Peninsula.[27] Despite the British service, shipping lines for the transportation of commodity and the communication of state administration between the islands in the Indies remained patchy.

Realizing the importance of modern transport infrastructure, the Dutch began to construct massive transport networks, facilities, and technologies in the 1880s. In 1887, the Stoomvart Maatschappij "Nederland" and the Rotterdam Lloyd agreed to merge and become the Royal Packet Navigation Company (Koninklijke Paketvaart-Maatschappij, KPM). The KPM sealed an exclusive contract with the government, which was signed in July 1888 and went into effect on January 1, 1891, to conduct regular mail and cargo services in the Indies. The KPM's exclusive monopoly in the Indies waters allowed the company to quickly develop a comprehensive shipping network to the exclusion of most of its rivals. Consequently, English liners became limited in their movements, which drastically lowered their profits in the Indies.[28]

The legal right for monopoly and government subsidy required the KPM to make several guarantees to the government. These guarantees both reflect their motives and interests and reveal their social and political implications. The requirements were as follows:

> The ships had to maintain a definite time schedule, transport the mails free of charge, establish maximum rates for passengers and for freight with special discounts for Government traffic; they had to be placed at the immediate disposal of Government in case of war, or else had to follow instructions issued to them by the commanders of Netherlands men-of-war in such events. Two thirds of the number of vessels to be built had to be constructed on Netherlands wharves.[29]

This excerpt urges the application of enlightenment ideas of order through standardization of time, rates, and routes. While indigenous peoples' traditional sailing ships were often used for colonial transports, their operations could not promise regularity, primarily because their operations were reliant on nature. In contrast, the modern, state-of-the-art steam-powered ships of the KPM were powerful enough to operate in almost any weather, realizing the enlightenment dream of progress in which machines tamed nature. A regular shipping schedule also transformed prevailing ideas of "time."[30] By standardizing time, spaces were compressed. Reaching an island was no longer about distance but rather about taking a standardized route and following a regularized schedule. In 1891, the distance between the many islands in the Indies and between the Indies and regions abroad was compressed into one regular shipping network that was facilitated by the KPM.[31]

The requirement also spoke to the state's needs to heighten bureaucratic administration. Mail was the method by which regular reports from government officials in smaller islands reached the central government in Buitenzorg in Java, so the priority of the KPM remained to serve administrative control and to facilitate military services. The KPM became the extension of the government in smaller scattered islands, and in a time of indigenous revolts, wars, and rebellions that often followed economic expansion, ships became the technological means to transport troops to restore public order.

KPM ships were also forced to use the Dutch wharves. After the establishment of the KPM, ports and wharves were constructed in the next decades, with Batavia, Semarang, and Surabaya serving as the three primary locations. In total, 300 ports, both traditional and modern, spanned the archipelago. The development of modern ports meant creating progress in communication networks. This progress was followed with the development of distinct types of vessels to fulfilling different geographical needs: freight, river boats, coal transport, and passenger.[32]

The implementation of the KPM served to modernize transport and to expand administrative control. Following the development of regular shipping lines, the colonial government saw the need to develop railways to connect port cities to the hinterlands. As early as 1840, discussions to construct a railway had begun because of problems that were related to transporting government produce grown under the Cultivation System in Java. The plan was hindered by the lack of familiarity with railways and rail construction, and hesitation around the capital-intensive nature of railway building. In 1860, a "Committee on Means of Transport" suggested that the construction of a railway from the port city Semarang to the hinterland would be profitable and would not expose any interested parties to financial risk. On August 10, 1867, the first part of the railway line between Semarang and the interior principalities was opened, built by Beyer, Peacock and Company in Manchester, and run by the Nederlandsch-Indische Spoorweg Maatschappij (Netherlands-Indies Railway Company).[33] From 1888 to 1925, railways, both belonging to the state and to private enterprise, had connected the island of Java. In 1888, the lines served only Batavia, Buitenzorg, and Bandung in West Java, while areas between Bandung and Semarang, northeast of Purwodadi, and east of Probolinggo were unconnected. In less than four decades, however, all of Java was connected by railway networks, which would soon allow for the development of a national consciousness and identity.[34]

The attention to railway and shipping networks demonstrated a larger colonial interest in transport communication. After the development of railways and shipping lines, the Dutch continued to create more advanced and modern transport facilities. Wim Ravesteijn notes that by 1950, 12,000 kilometers (7,456 miles) of asphalted surface, 7,500 kilometers (4,700 miles) of railways, bridges, irrigation systems—covering 1.4 million hectares (5,400 square miles) of rice fields—and several harbors had been constructed. Together they created the material base for an integrated Dutch colonial—and later Indonesian—state.[35] These transport systems united the archipelago and provided regular communication, facilitating the modernization of the Dutch colonial state in the Indies.

Technologies such as transport and communication infrastructures, however, are not neutral and apolitical; they are embodiments of social relations. The development of these structures cannot be separated from the labor that enabled their existence. The development of transport infrastructures and technology in the late nineteenth century owed to labor exploitation and genocide in the production of the preceding infrastructures. Toer provides a detailed account on the first development of the modern road that connected Java: Jalan Raya Pos (the Grote Postweg), also known as Jalan Daendels.[36] The road was named for Daendels, who, in 1808, experienced hardships after he arrived in the Indies. Daendels instructed the colonial government to mobilize corvée workers to build a 1,100-kilometer road from the far western part to the most eastern town of Java. To support both economic and military needs, the road had to follow military standards. For each 150,960-meter section, a pole was erected as a distance sign, and the road would have to be 7.5 meters in width.[37] Toer calls the development of Jalan Raya Pos the first "indirect genocide" in the Indies, as the construction of the road resulted in the deaths of over 12,000 workers.[38] Among the largest casualties occurred during the building of the roads connecting Buitenzorg and Karangsembung. About 1,100 corvées were deployed to build a road by breaking through high mountains with simple tools while eating a meager ration of rice with salt. Around 500 people died in Megamendung alone.[39] In other places, workers died of hunger, exhaustion, and malaria when they had to turn the swamps between Semarang and Demak into roads.[40] Toer notes that another genocide occurred during the Cultivation System, in which thousands of Javanese peasants and their family died of starvation. The forced labor led to the death of two-thirds of Demak's 336,000 population, and Grobogan's population decreased from 98,500 to 9,000.[41]

Major infrastructural developments occurred during the Liberal Period. The first was economic development, in which the traditional plantation model was industrialized and modernized through advanced technology and the science of production.[42] Starting in the 1880s, shipping lines operated regularly and linked the islands within the archipelago and connected them to the metropole, thanks to the opening of the Suez Canal in 1869. The construction of Tanjung Priok port in Batavia in 1870s, for example, boosted steam shipping activities, including tourism, between Europe and the East Indian archipelago.[43] Railroads and asphalt roads connected port cities with the interior regions of Java and Sumatra to ease the transportation of crops and commodities from the plantations in the hinterlands. Ulbe Bosma notes that while employment in the shipping industry in nineteenth through the early twentieth century skyrocketed, consistent and reliable statistics and information on the kind of labor involved in construction and infrastructure including roads, railways and irrigation works are scarce.[44] This absence points to the perceived unimportance of laborers by colonial authorities.

The Economy of an Archipelago

Although the implementation of regular and rapid transport systems in the Indies during the late nineteenth century was motivated partly by the need for more efficient and robust agricultural production, the implementation of regular and rapid transport systems opened new frontiers in the widely scattered islands in the Eastern part of the Indies. The construction of transport networks supported the economic and administrative pacification of this area.

The expansion of economic production required two commodities. First, more land belonging to the natives was, through government policies, forcibly taken and rented cheaply to private enterprises. Plots that were previously cultivated for the natives' livelihood and for planting various kinds of crops were converted into single-crop plantations—coffee, tea, sugar, coconut, rubber, and more. The following table shows the numbers of the total export production in Java and Madura as well as in the Outer Islands (Sumatra, Kalimantan, Sulawesi, Moluccas, and other islands) before and after the implementation of transport networks. Before the implementation of regular transport networks in 1888, production in the Outer Islands comprised just over a third of the production of crops in Java and Madura. By 1900, the export production in Java and in the Outer Islands increased by almost half, and by 1925, the export production of the Outer Islands exceeded that of Java and Madura. This relationship grew dialectically; as transport drove the growth of production, the expanding production created the need for new modes of transportation.

TABLE 1.   Total export in Dutch guilder fl.1,000

LOCATION	1879	1889	1900	1910	1925
Java & Madura	96,837	112,703	156,993	258,737	335,654
The Outer Islands	37,541	51,382	73593	153,658	453,784

Source: W. L. Korthals Altes, *Changing Economy in Indonesia: A Selection of Statistical Source Material from the Early 19th Century up to 1940*, ed. P. Boomgaard, vol. 12a, *General Trade Statistics 1822–1940* (Amsterdam: Royal Tropical Institute [KIT], 1991), 191–93.

Second, because of the increasing production and industrialization of production, more laborers were required. Bosma notes that during the implementation of the transport system, agricultural production was industrialized through the introduction of machinery and applied science.[45] The construction of more factories and plantations in the hinterlands and in remote islands required the availability of laborers, and transport networks transported workers en masse to these areas, enforcing massive migration of native population from one area to another. Previously, population movement between the various possessions within the Indies was limited by the Dutch colonial government. Toward the end of the nineteenth century, due to the demand for labor in the plantations in the east coast of Sumatra, the government permitted a massive flow of indentured laborers from Java to northern Sumatra, which was facilitated by the transport networks. According to the 1930 census, 443,000 of the people who lived on these plantations were born in a different administrative region.[46] The movement of people at the turn of the century from Java to the outer islands, as well as from densely populated central Java to the eastern part of Java that was populated with sugar plantations, both reveals the pursuit of colonial agricultural enterprise and explains the social background of the participants of the *pergerakan merah*.

By facilitating economic production and distribution as well as the mobility of labor, the implementation of transport technology and the industrialization and modernization of agricultural production was an unquestionable success for the colonial government. In the first quarter of the twentieth century, the Netherlands Indies enjoyed a healthy and successful export trade in agricultural produce to Europe, North America, and immediate neighbors in Asia. By 1925,

TABLE 2.   Destination and value of exports from the Netherlands Indies, 1928

DESTINATION	VALUE
Netherlands	f. 263 m
Other Europe	f. 312 m
United States	f. 201 m
Other America	f. 2.5 m
Australia	f. 50 m
Hongkong and China	f. 111 m
Japan and Taiwan	f. 57 m
Singapore and Penang	f. 326 m
India	f. 175 m
Other Asia	f. 31 m
Africa	f. 36 m

Source: Cribb, *Historical Atlas*, 143.

it was the second main exporter of sugar to the world after Cuba and the second highest sugar producer after India.[47]

While the regular shipping services of the KPM, as well as railway services, were deemed by the colonial government as a key element of economic development, they were also considered a symbol of the Dutch claim to sovereignty in remote regions of the Indies. The economic expansion to remote islands and areas in Java was followed by administrative spread: Local government offices were created and gave monthly reports that were shipped via the regular shipping lines.[48] This was reinforced by the recruitment of indigenous people—especially in Java—to work as government officials. Modern education that was available to upper class natives in the beginning of the twentieth century created a rank of the *prijaji* (read: *priyayi*, member of the Javanese official class), who mostly worked in local governmental offices.[49] The penetration of the Dutch colonial administration in remote areas in Java and the Outer Islands entailed the penetration of the state in the everyday lives of the people.

Despite the involvement of natives in running the colonial economy, the Liberal and the Ethical Policy periods failed to fortify the economic conditions of the Javanese. Wertheim argues that this is due to the process of cartelization. "The concentration of power and interest assumed ever larger proportions until, in the twenties, the plantations were all brought under a coordinating superstructure of large syndicates and cartels working in close cooperation with the government authorities."[50] The Liberal Period fostered the intrusion of large capital into the countryside, and created the conditions by which cartels could dominate not only the economy, but also colonial government policy and labor conditions.

Despite the economic expansion of the Liberal and Ethical Policy eras, the Dutch administration, including the organization of trade and commerce, was centralized in Java, creating a central and periphery relation, with Java being the center of state administration and of trade and commerce control. While economic production and state administration expanded from Java to the Outer Islands, economic distribution and state power became even more centralized. The transport networks facilitated this "centralization by decentralization" process by sending information—mails, reports, letters—to and from the Governor-General Office in Buitenzorg in Java and the rest of the Indies.

Colonial Social Networks

As the growth of networks of mobility enabled the movements of people between the islands and between the Indies and Europe, so too did social networks of the press, postal service, and the telegraph. Before the mail steamers provided by the KPM in the late nineteenth century began to facilitate the circulation of governmental reports and letters between islands, the VOC likely had begun the same service on a smaller scale. The printing press was also brought by the Dutch merchant company to the Indies, but its use was limited to VOC-related matters. Newspapers were first printed as government's means of mass communication and Dutch missionaries' means of proselytization, but in the late nineteenth century their function shifted and they became tools to support commerce through advertisement and to educate people in the science and technology of agriculture.[51] The postal service was incorporated as a government service in 1862, and the telegraph, which was developed much later, also created a pivotal change in the circulation of information.[52]

Throughout the late nineteenth century until World War I, the telegraphy service of the Dutch empire depended on the British submarine cable that connected Asia and Europe. By 1900, the British had linked all parts of their empire with telegraph cables often called the "all red line." With this line, a message from India could reach London within thirty minutes whereas a letter would have taken weeks. However, "the all red line was a highly controlled media environment, operated by monopolists that charged high rates per word, which meant that many people could not even afford to send telegrams."[53] Intercontinental telegraph lines were monopolized by three large press agencies: Reuters (British), Havas (French), and Wolff (German). Not possessing independent telegraph lines meant that Dutch press agencies had to rely on news that was provided by these monopolists.[54]

Access to this undersea telegraph cable network halted during the war. Despite its neutrality, the Netherlands' dependence on foreign powers for its colonial telegraph connection affected its communication with its colonies. In all Dutch ports, both in Europe and Asia, a blockade was imposed by the British fleet. Any vessels coming in and out of the harbors were stopped, Dutch ships carrying mail were banned, and the telegraph stopped running.[55] Consultations between the colonies and the metropole became impossible.[56] As a result, the Indies suffered tremendous economic and political problems. Anticolonial struggles within the Indies became stronger and found an inspiration from the Russian Revolution that soon provided a platform for solidarity among the colonies.

During this period of isolation, the minister of colonies hired C. J. de Groot, who had just defended a PhD thesis at the University of Delft on long-wave transmissions in the tropics. In the thesis he found that a direct radio connection between the Netherlands and its colonies was "political necessity and technically possible."[57] This work provided hope for the government, as previous efforts to connect the 12,000-kilometer distance between the Netherlands and the Indies had been in vain. Aided by a fl.5,000,000 grant, De Groot arrived in the Indies in 1917 and was tasked with putting his theory into practice. Tropical conditions proved to be a major challenge. To get a performance comparable to cables, for example, "one required at least six to eight times more radiated power than under comparable European conditions."[58] De Groot decided that substantial machinery would be required to establish long distance radio connections for long-wave technology. To that end, he built a high-powered station at the Malabar Gorge in the mountains surrounding Bandung in West Java.[59] With materials smuggled from Japan, the development of the station required him to modify the transmitter that would boost up the transmission power of "what would become the most powerful station on the earth at the time."[60]

As efforts to build the telegraph infrastructure increased during and after the war, D. W. Berretty, founder of a press agency in Java called Algemeen Nieuws-en Telegraaf-Agentschap (Aneta), struck a deal with Reuters and bought its office in Batavia. It provided him with a monopoly over the distribution of foreign news in the Indies and news about the Indies in the Netherlands.[61] When communism became popular in the Indies, the Dutch authorities tightened their ties with Aneta and other press agencies via increased funding. Aneta received fl.15,000 from the Departments of Foreign Affairs and Colonies' secret funds. A colonial entrepreneur added another fl.20,000, and Aneta's taxes were reduced by the government. Vincent Kuitenbrouwer argues that "a three-pronged colonial media strategy emerged that aimed to control the telegraph lines from the Indies, increase the flow [and distribution] of official publications . . . directly to journalists via the information bureaus and embassies."[62] The communist uprisings of 1926 to 1927 became a catalyst for a coordinated effort by journalists and press agencies, in direct contact with government officials, to manipulate international public opinion on the popular communist movement in support of the colonial regime in the archipelago.[63] Social networks of communication like newspapers and the telegraph helped control and manage the worker population by reproducing the colonial social structure of inequality based on race, class, and political orientations. In other words, social and cultural organizations helped organize the economic production of the colony. This also includes the production of colonial maps.

Spatial Fantasy and Transport as Workplace

Dutch modernization projects utilized cartography to conquer, subject, pacify, and manage the colonial population. Maps became the avenue for colonial power to visually represent its geographical rule and to fulfill its need for rhetorical self-creation. Through maps, colonialism was disguised behind the discourse of progress, and the unequal hierarchy between the islands and between the people was masked in the idea of a "united" archipelago. Indeed, previous scholars have suggested how maps were part of the making of power.[64] Such maps became a part of formal education.

In the beginning of the twentieth century, when the colonial government began to enjoy benefits from its "progress," its pride was communicated by putting atlases in the classrooms. In 1920, the "Java Productenkaart" (Java Production Map) started to appear in schools. This map later became a part of the atlas the "Wandkaart van Nederl.-Oost-Indië" (Wall Maps of the Netherlands East Indies) by W. van Gelder and C. Lekkerkerker, which were produced in 1928.[65] The map also highlighted economic production and the network of transportation. It pinpointed locations of rice, wood, coconut, sugar, coffee, tea, kina, rubber, cacao, salt, petroleum, tobacco, and pepper production across Java. With these maps, students were exposed to a visual means through which they imagined the spatial construct of the Indies using the framework of economic productivity.

Maps also eventually pushed forward a new form of mobility: tourism. As the tourism industry boomed in the early twentieth century, thanks to regular shipping lines and maritime technology that reduced travel time from the Netherlands to the Indies to only a few weeks (compared to six to eight months two centuries before), the railway system became an important aspect of the map produced by the Official Tourism Bureau in Rijswijk 18, Batavia, in 1923. The map shows the geographical imagination of exotic places in Java and the available modern railways that could take people to tea plantations, mountainous areas, and hot springs in the hinterlands. Even De Groot's Malabar station, the symbol of colonial engineering, became an attraction for tourists who wanted "to marvel at the wonders of Dutch radio technology."[66] These maps represent the Indies as powerful, productive, and modern. For example, written in English, the map juxtaposed "places of interest for tourists," such as lakes, rivers, mountains, and Hindu temples and ruins (Java was once one of the biggest Hindu kingdoms) with detailed modern networks of "mailsteamers," "railroad and tunnel," "steam tram," "high and cart road," as well as "horse and foot path." Elsewhere, tea plantations were also advertised as objects of tourism, boasting the Indies' productivity and economic power.[67] The construction of transport networks in the Indies as modern yet exotic in the tourism discourse, however, conceals the exploitative aspects of colonialism.

The true contradiction of colonialism is revealed once transport is understood as a workplace. Tourist brochures circulated during the period juxtapose tourists on board with Javanese seamen working as servants on board the *Koninklijke Rotterdamsche Lloyd* (KRL) ship to the Indies.[68] Native Indonesians were hired in the position of lower-class seamen, working as servants. They were required to wear traditional clothes to maintain their indigenous identity. The photo boasts how the exotic natives will serve the white European tourists as they arrive in the

Indies, exploiting the colonialist imagination of submissive natives. The lighting, composition, and cosmetic look of the passengers indicate that this photo was staged and taken in a studio. The contrasting status between the indigenous Indonesians in uniform and the white tourists in fashionable clothing was purposefully crafted and reproduced to symbolize colonial success. In turn, it shows how the taste of tourists' attraction was shaped.

The official maps reveal how colonialism was ideologically produced and preserved by concealing the very exploitative aspect of colonialism itself. The enlightenment idea of progress underpinned the technological transformation of the Indies, especially in the second half of the 1800s. The next section will focus on the spatial changes in the colonial geography in the long nineteenth century and explore how these changes transformed ordinary forms of mobility and sociality.

Native Class Structure and Their Mobility and Sociality

Although the mobility and sociality of native Indonesians were transformed by the colonial state, indigenous populations moved around extensively prior to the arrival of modern steamships. In fact, as Bosma argues, the "image of Java with self-contained villages that were devoid of social diversity and social mobility was a colonial perception rather than a traditional feature of Javanese society."[69] Before the arrival of Western merchants, the archipelago was bustling, frequented with large indigenous ships facilitating large-scale trades of produce such as rice and peppers between the islands.[70] Like Europeans in the Atlantic economy, the Indian Ocean and the Pacific were filled with Asian migrants who traveled to different areas due to economic stimuli and labor demands.[71] The precolonial mobility for trades and exchanges inter-Asia and between the islands in the archipelago was dominated with the mobility of labors.

The world of VOC was "a world of labor," and it required large numbers of maritime workers.[72] It recruited many European and Asian laborers, including unfree laborers or slaves and free and semi-free laborers from different regions in Asia.[73] Among them were Moorish sailors from Bengal, Surat, and other Indian ports. Over time, the Moors who worked in and around the port of Batavia settled there and through social interaction, they changed the demographic makeup of the city.[74] Globalization circa 1750 to 1850 also allowed for the circulation and mass consumption of Southeast Asian products in China. During this time, Chinese migrants, along with Indonesian and other Asian migrants, flooded the Malay Archipelago "in search of Marine and forest products, better protection from local rulers, and better locations for their trade and maritime raids."[75]

At the end of the eighteenth century, as Dutch colonialism transitioned from old European mercantile colonialism to the new land-based colonialism, the colonial government sought to control the movements of Javanese rural population.[76] In order to accrue capital, colonial policies exploited forced labor, restricted the movement of people, and kept peasants bound to their land (later colonial policies similarly restricted mobility as a response to the communist movement).[77] However, the Cultivation System proved to be burdensome, and it triggered migrations among the local Javanese. At the end of the Cultivation System, as the colonial state was modernized, mass migration continued due to the emergence of hundreds of plantations, transport, construction, and urbanization including Indonesians, Chinese, Arabs, and Europeans to and from Java and other islands.[78]

The Cultivation System transformed the economy from subsistence agriculture to commercial agriculture, and most of the natives were forced to cultivate crops for export. Robert Elson argues that this firmly set the economy from what used to be an age of commerce to an age of peasantry.[79] The expansion of land use for plantation agriculture continued to expand through to the twentieth century, making Netherlands East Indies distinguishable from most of the rest of Southeast Asia. The majority of the natives became occupied in peasantry, which was later crucial for the rise of the Indonesian Communist Party (PKI).[80]

Additionally, export production also necessitated transporting produced crops from the hinterland to the ships at the coasts. This created new professions outside of peasantry: porters, coolies, loaders, escorts, and messengers, as well as peasants who provided their horses, buffaloes, carts, or traditional ships in waterways. These professions served the business of transporting, and corresponding, between areas. With the modernization of communication networks in the subsequent decades, these professions would only expand. The surplus of agriculture led to new opportunities for the natives to work in *keris* (Javanese traditional dagger) making, jewelry making, and traders, which had previously been dominated by the Chinese.[81]

These changes in occupation reveal the emergence of what Radin Fernando calls the "non-agricultural economic activities."[82] Even though during the Cultivation System the mobility of peasants was restricted, the natives who held nonagricultural professions were able to move around. James Rush shows the existence of peddlers, theatrical players, musicians, and smugglers who were as mobile as transport workers and traders.[83] These workers penetrated Java and created new sociability as *pasar* (traditional market) and *warung* (stalls) emerged along the roads. In these markets, imported products from China and the West were sought after along with local products. In the past, Chinese traders "took the goods imported from the Netherlands and other European countries to the interior and brought back the products of the interior to colonial towns and harbors";

however, as natives became more involved in nonagricultural economic activities, more and more of them became traders.[84] Nonagricultural economic activities were usually compensated with money, showing the beginning of money as means of exchange. This transformation brought about by the non-agricultural economic activities expanded especially during and after the Liberal Period.

The expansion of the colonial state and the modernization of its infrastructures toward the end of the nineteenth century through the early twentieth century transformed ordinary Indonesians' mobility and sociability. At this time, the peasantry was pushed into cities and developed into a proletariat. A large number of peasants became landless and had to find jobs elsewhere, which led to the process of the proletarianization of the peasantry. Workers migrated to plantations within Java and to other islands like Sumatra. Other wage laborers moved to the cities for nonagricultural economic activities. They were able to receive modern education and training. In general, labor relations outside of agrarian activities, like mining, industry, and transport, are the least traditional, and these workers were exposed to modern organizational structure and higher wages. "The modern environment of mining, transport and industry inevitably affected the attitudes of workers as well."[85] These were the lower-class ordinary people who would later mobilize the movement.

The new forms of mobility and sociality transformed the daily living conditions of ordinary subjects. During the time of the *pergerakan merah*—some thirty years after the Dutch colonial government began to modernize transports—railway services had become an ordinary part of life for the indigenous people living in the cities and in the hinterlands. By 1920, transport networks had connected urban and remote areas in Java and the Outer Islands. The number of railroad passengers in the Indies in 1929 indicate the mundanity of transport.[86] Roughly 126,259,075 indigenous people were using railway services in Java that year. If the indigenous population in Java was 37,393,740 people, then one of them would use railway services 3.4 times on average.[87] At first glance, this number appears to be small, but in comparison to the condition of natives' mobility just sixty years before, it represents a massive leap forward.

Before the railroad was installed, the main transport was a horse-drawn carriage, which was accessible only by upper class people. The lower class's mobility was largely limited to travel by foot and, occasionally, horses. Multatuli's description in his 1860 work *Max Havelaar* offers an example of this poor and often wretched condition of mobility in Java half a century after Jalan Raya Pos—the Great Road—was first built,

> "The main road" is perhaps a slight exaggeration in respect to the wide footpath that ... one called the "road." But when, with a coach and four,

one started for *Serang,* the chief township in the residency of *Bantam,* intending to drive to *Rang-Betoong,* the new center of *Lebak,* one might be fairly sure of arriving there some time or other. It was, therefore, a road. It is true that time after time one would be stuck in the mud, which in the *Bantam* lowlands is heavy, clayey, and sticky . . . The driver would crack his whip, the "runners" [or footmen, people who helped a coach from getting stuck in the mud] . . . with their short thick whips, trotted again by the side of the four horses, shrieked indescribable sounds, and beat the horses under the stomach by way of encouragement.[88]

Comparing conditions of mobility before and after the availability of modern transport explains how material conditions affects the organization of resistance movements. Limited mobility meant that opposition was sporadic, local, and traditional, as noted in Robert Cribb's map of "colonial warfare and indigenous resistance, 1815 and 1910."[89] Though resistance against colonial power frequently occurred across the Indies archipelago between 1815 and 1910 including the Java War and the Aceh War, these resistance movements were isolated from each other. During the *pergerakan merah* of the 1920s, however, public transportation was already in place, connecting most areas of Java, making other cities, towns, and villages accessible for ordinary people.

Motivated by the evolving interpretations of enlightenment adopted by Dutch officials, the development and modernization of transport networks and infrastructures explains the national, archipelagic, and global nature of the *pergerakan merah*. The *pergerakan merah* produced a red enlightenment—a particular understanding of enlightenment that motivated their version of revolutionary anticolonial struggles and the revolutionary communication they produced. This is why the *pergerakan merah* is best understood as an event of communication. The red enlightenment emerged within specific contours of communication networks and was produced by way of repurposing colonial communication into revolutionary communication.

Communication Workers at the Forefront

The novelty of the *pergerakan merah* came from the centrality of the activities of borrowing, adopting, and repurposing the existing communicative means of colonialism into those of resistance. The creative energy of the revolutionaries was acquired partly through the nature of their own occupation. While communication networks did facilitate and expand the reach of connection and mobility among ordinary people, communication workers took the greatest advantage of this material condition and eventually organized the *pergerakan merah*. They were highly educated, better paid, and highly skilled, and had access to information and mobility. Modern education and training, exposure to the exploitative nature of capitalist economy, and contacts with modern organizational structure, as well as workers of diverse backgrounds, shaped their particular class consciousness. Sartono Kartodirjo argues that Sarekat Islam (Islamic Union, henceforth SI, one of the main organizations leading *pergerakan merah*) was a modern organization led by educated workers in the cities.[90]

This section shows the communication networks in the Indies from the perspective of the communication workers and their effort to mobilize the movement through the activities of repurposing. These workers envisioned to democratize access to communication and to build an emancipatory movement for liberation. By the 1920s, four biggest unions in the areas of communication were already working under the grasp of the communist umbrella: Sarekat Postel (the Union for Post, Telegraph, and Telephone workers, SP), Vereeniging voor Spoor-en Tramweg Personeel (the Union for Railway and Tram Workers, VSTP), Sarekat Pegawai Laoet Indonesia (Union of Indonesian Seamen, SPLI), and Sarekat Pegawei Pelabuhan dan Lautan (Seamen's and Dockers' Union, SPPL).

With direct access and control to the means of sociability, workers in the Post-, Telegraaf-, en Telefoondienst (PTT) became involved in the day-to-day expansion and communication of the *pergerakan merah*. SP (previously Karoekoenan Pegawai Post Boemipoetera [the Union for Indigenous Postal Personnel]) was founded in the early twentieth century and was headquartered on Karreweg 42 at the center of the radical city of Semarang. The group worked closely with the leadership of other communist affiliated organizations, the PKI, SI, and VSTP. On the board of SP was Sudibio, Alibasah, Samsi, Mohamad Tahir, and Kadarisman, and its main organ *Soeara postel* (*The Voice of Postel*) was led by Soekindar, Soebakat, Soekartono, and Soepeno, all well-known communist leaders who also organized other communist unions or organizations. Outside Semarang, SP's branches in Sukabumi and Bogor were also led by communists. The communist Sardjono, for example, was the main editor of the *Soeara kita* newspaper, which was published by SP's West Java branches.[91] The active roles of these communist leaders in SP reflects an affinity between PTT workers and the *pergerakan merah*. While the communists saw the need to expand its leadership in this state-owned communication service institution, PTT workers preferred to choose union organizers who were not PTT personnel—people who had the time and the energy "to freely administer and visit *vergaderingen* in other places."[92] The Semarang activists were natural choices, especially given the fact that the same activists worked closely with VSTP leadership who had well-known success in winning workers' demands.

Having SP under the influence of communism was beneficial because they controlled the day-to-day operation of social network in the Indies. While all civil servants and officials working at the PTT could become members as long as they paid their dues, most of SP's paying members were lower-ranked officials who worked as *brievenbestellers* (postmen), mail sorters, telephone mechanics, telephonists, and clerks—people who worked directly with the daily operation of message and mail circulation. Their participation allowed the *pergerakan merah* to encourage its members to smuggle letters and publications free of charge.[93] SP's network gave promise for information to be transferred smoothly at a low cost, given that it had reported membership of over 1,700 in 1923, 50 percent of whom were lower-ranked officials, with branches in multiple cities across the Indies, Cirebon, Bandung, Sukabumi, and Purwakarta in West Java; Madiun, Solo, Semarang, Pekalongan, Kudus, Tegal, and Purwokerto in Central Java; Bondowoso, Banyuwangi, Probolinggo, Malang, Situbondo, and Wlingi in East Java; and, outside of Java, East Sumatra, Medan, and Aceh.[94] This communist infrastructure of information that repurposed existing official network of communication became an integral part for movement building.

PTT workers were willing to risk their jobs in the name of *kemanoesiaan* (humanity). They believed that these risks served a greater good.

> SP is not meant to incite crime, but rather to further the movement. You need to be ensured that you believe in yourselves. I am worried that you will be hopeless thinking that what we experience is just our fate, that cannot be changed with man's efforts. This means you don't remember humanity. You should know that humanity is sacred. . . . [But] we cannot keep this sacredness alone. We should rather unite with our fellow workers to loosen the grasp of our enemy's hand.[95]

Using the religious reference of sacredness, the writer discusses the goal of the movement: *kemanoesiaan* (humanity). He argues that this goal tramples a belief promoted by religions that human life is a given and that suffering is fate. By changing this superstitious belief into a belief in humanity, the writer contends that one is then capable of changing his/her circumstances. This sacred effort, however, is not an individual task, but it rather should be done collectively in solidarity with other fellow workers.

These spirited efforts to risk one's life and organize in the movement frightened the colonial state. The PTT's archives contained numerous stories of surveillance and coordination between the police and PTT higher management. As early as 1923, the police closely monitored the activities of SP. They often hired other workers in the PTT to work as spies to keep watch on their fellow workers. Police reports would be circulated to the higher ups of the PTT, warning them of the involvement of their workers in communism, and urging them to fire the communist workers. (Their suspicions were usually based on the reported lists of attendees, speakers, and comments made in communist meetings.)[96]

Publication of communist works was also highly controlled. In 1926, a copy of *Soeara postel* was seized while en route to the print shop in Solo because the police believed it contained articles that incited hatred against the postal service. By this time, the headquarters of SP and the editorial board of its organ had been moved to Surabaya, about 200 miles east of Semarang, but they used the Merdika printing office in Solo, which is just sixty miles south of Semarang.[97] Only a few printers were willing to produce publications that expressed revolutionary views; "other *drukkerij* [printing houses] refused due to its *revolutionair* [revolutionary] views."[98]

While SP mobilized post, telegraph, and telephone workers, VSTP and SPLI organized transport workers. Like SP, they both had large membership and had branches all over the Indies and abroad. VSTP was founded on November 14, 1908, and, under the leadership of the young Indonesian communist Semaoen, it became bigger and more successful.[99] Railway workers working in different railways companies, such as Staatsspoorwegen (SS), Nederlands-Indische Spoorweg (NIS), Semarang-Cheribon Stoomtram Maatschappij (SCS), Serajoedal Stoomtram Maatschappij (SDS), Oost-Java Stoomtram Maatschappij (OJS), and Semarang-Djoewana Stoomtram Maatschappij (SJS), as well as other railway related services, all enthusiastically enrolled in the VSTP, reaching a peak in October 1921 at just under 17,000 members. Most of the members, like SP, were Boemipoetera (Indonesians) and lower ranked officials, with some Dutch and Tionghoa (Chinese) members.[100] The spread of VSTP branches during those years was also rapid, averaging about thirty new branches annually. While a couple of them met their demise, by the end of 1922, VSTP stayed strong, operating 146 active branches. During the mass VSTP strike of 1923, which will be discussed further in chapter 3, the membership rose back up to 13,000.[101]

TABLE 3. Number of VSTP membership in 1920–22 by race

YEAR	BOEMIPOETERA	DUTCH	TIONGHOA	TOTAL
Beginning of 1920	6,235	236	23	6,494
End of 1920	12,084	95	34	12,213
Oct 1921	16,831	104	40	16,975
End of 1921	15,621	102	46	15,769
June 1922	7,642	45	44	7,731
End of 1922	9,549	43	15	9,607

Source: Hoofdbestuur, *Poesaka V.S.T.P.*, 144.

The publication of VSTP's organ *Si tetap* was not without challenge. Much like SP's *Soeara postel*, it was often boycotted by capitalist printing houses. As a result, *Si tetap* had to borrow the facility that was being used by SI to print its own publication *Sinar Hindia*. In 1922, however, VSTP was able to raise fl.3,000 to buy its own printing press. Using this machine, VSTP's publication of *Si tetap* and *De volharding* became less costly. As capitalist renters did not want to have anything to do with the revolutionary union, VSTP was forced to buy an office space at Heerenstraat, Semarang, for fl.24,000.[102] These purchases allowed VSTP to spread its wings and became one of the leading organizations in the *pergerakan merah*.

SPLI and SPPL were the other two large unions that organized communication workers. In its first year in 1924, over 1,200 Indonesian sailors became members of SPLI, and by 1925, SPLI's Indonesia-based organization, SPPL, united all of the existing seamen's and dockers' unions into one organization. The story of SPLI and SPPL will be discussed further in chapter 7.

With members covering strategic positions that managed and controlled communication infrastructure, the movement was able to repurpose formal communication networks into means of movement building. PTT workers, railway workers, and sailors connected ordinary people in different places and created a network for mobility and sociability, enabling the movement to reach beyond one location to most areas in Java, in the Outer Islands, and outside of the Indies globally.

A Taste of Their Own Medicine

Modern transport networks in the Dutch East Indies that were meant to facilitate the increase of colonial power became one of the sources of colonialism's failure. While the networks helped the spread of the economic, political, and social conquest by the colonial state, they also created conditions that would lead to the spread of anticolonial resistance. Transport networks made possible the increased mobility of indigenous people as labors and transport workers, which helped create new forms of contacts and associations that allowed global and transisland and transregion coordination. Transport networks also made physical mobility a part of the mundane life of the ordinary indigenous people. This new form of mobility allowed the creation of collective communities beyond geographical locations. This explains the national, archipelagic, and global scope of the movement.

The cosmopolitan and global nature of the *pergerakan merah* was the result of the evolving enlightenment aspirations that influenced colonial policies around the development of communication networks and infrastructures. The different interpretations of enlightenment by Dutch officials are key to understanding the broader context of the development of modern communication networks, the changes of natives' class structures, and the emergence of communication workers as engines of the movement. Enlightenment was understood by Dutch officials more as civilization, order, and progress than social justice and liberation. The liberal versions of enlightenment these officials brought to the Indies did not free the native population. Instead, they strengthened the chains. The modern education the upper-class Javanese, *prijaji*, received gave them an avenue to be absorbed into the colonial state structure. In turn, this separated them from the lower-class Javanese, *kromo*, who continued to be deprived of education. The proletarianization of peasantry introduced natives into a wage system and furthered urbanization process, creating kampongs (poor pockets in cities). The result is a class contradiction and its entailing cultures: the educated versus illiterate, modern city life with European goods versus dirty kampong, and a racial contradiction between natives, Europeans, Chinese, Arabs, and other Asians. The emancipation promised by the liberal era was only enjoyed by a segment of the native population. It took communism and its discourse of class to spark the mobilization of universal emancipation by lower class people to fight against colonial injustice and liberation. In the next chapter, we will delve into how the *pergerakan merah* in 1920s borrowed, modified, and took upon enlightenment rhetoric, and combined it with communist rhetoric, to build their emancipatory project.

RED ENLIGHTENMENT IN THE ROARIN' TWENTIES

The expansion of communication networks transformed social lives in the Dutch East Indies. Contacts with the Dutch introduced natives to new developments in paper making, book making, book binding, and printing technology, leading to a literary awakening in the latter half of the eighteenth century that Theodore Pigeaud calls a period of renaissance. Tales that exalted the kings began to be recorded in Javanese scripts in modern book forms, and the ease of mechanical reproduction sparked a burst of literary creativity. At the center of this activity were traditional intellectuals called *pujangga*, the custodians of the great tradition of Javanese royal palaces who worked under the patronage of the kings. Their writings were preserved in the palaces as sacred regalia.[1] Kenji Tsuchiya, in chronicling the story of a *pujangga* named Ranggawarsita, argues that this transformation of Javanese literature did not occur in isolation; he argues that the literary renaissance existed due to direct contact with Dutch academic world and the global expansion of modern printing. As Tsuchiya puts it, "the modern age of Javanology touched the Central Java kraton [palace] long before the twentieth century began."[2]

In the latter half of the nineteenth century through to the fin-de-siècle, access to print production and consumption trickled down to people of half-Dutch and half-indigenous origin, wealthy Chinese merchants, and native aristocratic communities. Together they gave birth to the colonial lifestyle that was "filled with mimosa and bougainvillea," where literature and the arts flourished.[3] Eurasians contributed to the mestizo culture by advancing *keroncong* music, which originated from a Portuguese musical tradition, and the "beautiful Indies" genre of landscape painting.[4] Didi Kwartanada writes that enlightened peranakan Chinese women participated in classical music, vocal and piano recitals, ballet, and ballroom dance, and their families adopted French names.[5] An educated class of natives, *kaum terpelajar* (the educated), or what William Frederick calls the "new *prijaji*" thus emerged. Unlike the previous *prijaji* class, which received its elite status based on hereditary and aristocratic lineage, the new *prijaji* ascended by acquiring western education. Many of them served as government employees as the state expanded its offices and penetrated rural areas and other islands outside of Java.

During the roarin' twenties, a cosmopolitan society was fostered through cultural interaction. Pastiche, imitation, adaptation, and consumption of foreign arts became cornerstones of the era. Vernacular customs and foreign products were so enmeshed that one could not discern where one ended and the other began. Komedi Stamboel, a theatrical performance that was accompanied by singing, instrumentals, and dance, was introduced in Surabaya in 1900s, and helped develop and popularize *keroncong* songs.[6] Using gramophone recordings, Stamboel appropriated foreign music from Arabic music to jazz and popularized its appropriation. Exposed to jazz by way of the Filipinos' "Manilla jazz," Stamboel blended *keroncong* with jazz, tango, and rumba, and created its hybrid forms of "*keroncong* tango" and "*keroncong* blues," collectively known in its popular name "jolly jasz."[7] *Keroncong* quickly became the dominant popular music genre throughout Java and, to an extent, the entire colony, and birthed a social dancing and a celebrity culture that produced the first native female star—a singer named Miss Riboet.[8]

The period of *keroncong*, Komedi Stamboel, and cinema differed from the literary renaissance and the mestizo culture in that the cosmopolitan culture was available to commoners.[9] The production of ideas and culture was no longer centralized in the hands of the elites like Ranggawarsita and had moved away from royal regalia and religious texts. Middle-class ideas of emancipation and the urban consumer-oriented lifestyle were displayed in mundane places like newspapers, *pasar malam* (night markets), and theaters. This structure of feeling is captured well in the short story "Aliran Djaman," which takes place in a night market, a *societeit* (an elite social club), and a *vergadering* (meeting) place.[10] The *clubgebouw* (clubhouse) "Panggoegah" that were frequented by the young revolutionaries in the story was equipped with a library that housed newspapers and Western literature, a room for *strijkorkest* (string orchestra) where men and women danced and sang international revolutionary songs, and a sport room where players spoke Dutch, Malay, and Javanese. These spaces reflect how the revolutionary time became a novel, exciting, and alluring period of cosmopolitan culture in which "*lampoe elestrisch* [electric lightbulb]" symbolized a new era of hope and excitement for ordinary natives.[11]

This sense of modernity shaped ideas of emancipation among Indonesians. Likewise, enlightenment aspirations shaped the emancipatory spirit among Indonesian natives. These aspirations, combined with the circulating new communist ideas, evolved into the red enlightenment that motivated the production of communicative techniques and strategies of mobilization. In 1918, the conclusion of the war had brought food shortages, high costs of living, droughts, an influenza pandemic, and riots and strikes, and the poor suffered widened economic gaps and lower standards of living.[12] By the 1920s, the language of awakening, liberation, and progress, as well as that of radical communism, seeped into everyday lives. Previous scholars have rightly characterized the turn of the century in terms of modernity, but questions about the roots of the *pergerakan merah*'s revolutionary ideas have remained unanswered.[13]

This chapter is divided into two sections. First, it analyzes the evolution of enlightenment ideas that motivated anticolonial resistance, from Multatuli to Kartini to new *prijaji* figures and organizations. In the second section, it delves into how commoners took upon these ideas from the native elites and transformed them into a red enlightenment. While the enlightenment offered the language of emancipation, communism turned the circulating emancipatory spirit into radical anticolonial resistance. Communism helped hand over the language of emancipation from the *prijaji* to the *kromo*. This *kromo* version of enlightenment is reflected in both verbal and symbolic speeches that were demonstrated in slogans, cartoons, and feuilletons.

An Arc of Enlightenment Lost in Translation

Enlightenment-inspired critiques against Dutch oppressive practices in its colonies were voiced as early as the turn of the nineteenth century by the van Hogendorps, but it was Edward Douwes Dekker, writing under the pseudonym Multatuli, whose seminal work *Max Havelaar* (1859) sparked a sustained campaign for the emancipation of natives. For over two decades beginning in 1838, Dekker held various colonial government posts in the Indies, which gave him firsthand knowledge of the oppressive exploitation of Dutch empire. These experiences helped shape his thinking when he later became a writer upon his return to the Netherlands. Dekker had a strong belief in Enlightenment ideals and a Romantic sensibility. His idea that "people should feel solidarity with their fellow men because they share the same essence" resonate with those of Baruch Spinoza, a seminal Dutch Enlightenment thinker in the seventeenth century.[14] Carl Niekerk observes that Dekker is a defender of radical Enlightenment and had great admiration for Napoleon Bonaparte, M. de Voltaire, and Jean-Jacques

Rousseau.[15] He was a regular contributor to *De dageraad*, the radical magazine founded in 1856 to propagate Enlightenment ideas and the materialist and atheist concepts of the nineteenth century.[16] His novel *Max Havelaar* had remarkable impact in the development of Western scholarship, and the rise of large and powerful socialist parties in German-speaking countries sparked a renewed interest in *Max Havelaar* in the early twentieth century, due to heightened Marxist critiques of European imperialism and colonialism.[17]

In the Indies, *Max Havelaar* inspired a young Javanese princess named Kartini, who set out to be the first modern "Indonesian thinker."[18] As she grew up, Kartini enjoyed the cosmopolitan culture of the late 1800s. Facilitated by regular shipping, allowing mail, books, and reading materials to circulate in the Indies, Kartini met various Dutch officials and absorbed Western literature, especially during her time of *pingit* (seclusion)—a common practice among Javanese nobles that kept girls at home from age twelve through the time of their marriages. The suffering and confinement of *pingit* led to Kartini's interest to enlightenment. "While trapped in the house, she devoured books and newspapers, she listened to her father's conversations, she scrutinized the behavior of his guests."[19] Kartini shared her thoughts and observations in her correspondences with her Dutch colleagues and, after her untimely passing, a collection of these letters was published in the original Dutch language titled *Door duisternis tot licht* (*From darkness to light*, 1911).

Kartini's writing was the first writing by an indigenous Indonesian ever recorded discussing modern ideas of emancipation. The collection shows a strong influence of enlightenment ideas, and Kartini frequently uses the vocabulary of enlightenment in her letters. The word "*beschaving*" (civilization) appears fifty-nine times, "*ontwikkeling*" (development) sixty-four, "*vrij*" (free) 133, "*verlicht*" (light) fourteen, and "*emancipatie*" (emancipation) three times. Many scholars—even contemporary accounts—of Kartini often ascribe the word "modern" and "nation" to her thoughts, but this is inaccurate; the word "modern" only appears eight times and "nation" only three times. These readings deemphasize Kartini's aspirations for modern education for native children. For Kartini, modern equals enlightenment.

Kartini's letters were translated into English in 1921 under the title *Letters of a Javanese Princess* by Agnes Louise Symmers. The Malay version is titled *Habis gelap terbitlah terang*, and was translated by Empat Saudara and published by Balai Poestaka in 1922. In the English version, the words "*beschaving*," "*ontwikkeling*," and "*verlichting*" are interchangeably translated into "enlightenment," "civilization," and "education." However, the Malay translation of these three words uses copious expressions including "*berkesopanan*" (to be civil/ ethical/polite), "*kehormatan*" (honor), "*kepandaian*" (intelligence), "*kemadjoean*" (progress), "*berpengetahoean*" (knowledgeable), "*pendidikan*" (education), "*terang/memantjar*" (bright/light), and "*ilmoe pengetahoean*" (science). The word enlightenment, which in Dutch has the root "*verlicht-*," is translated in Malay into "*pendidikan*," "*berkesopanan*," "*ilmoe pengetahoean*," and "*terang*."[20]

A more in-depth study on these translations is needed to provide a comprehensive account on the subject, but this brief survey helps us unearth a couple of important findings. First, enlightenment ideas were very much alive in the Indies, especially among educated natives. Kartini read and educated herself in Western literature, and this knowledge is reflected in the quotes she refers to in her letters, including from *Max Havelaar* and from other literature on French revolution and women's movements in Europe. These books demonstrate her cosmopolitan influences. Second, the Malay version of enlightenment vocabulary shows that the traces of enlightenment ideas in her thinking were misinterpreted. The ways that the Dutch vocabulary of enlightenment was translated in the Malay edition show that "*berkesopanan*," "*kehormatan*," "*kepandaian*," "*kemadjoean*," "*berpengetahoean*," "*pendidikan*," "*terang/memantjar*," and "*ilmoe pengetahoean*" alluded to enlightenment ideas and aspirations. However, these terms were lost in translation due to a dominant nationalist framework that emphasized terms like "modern" and "nation."[21] As we have seen, enlightenment ideas are not limited to the transposition of the metaphor light or darkness as the late Southeast Asian historian Benedict Anderson suggests, but can be traced in "*berkesopanan*," "*kehormatan*," "*kepandaian*," "*kemadjoean*," "*berpengetahoean*," "*pendidikan*," and "*ilmoe pengetahoean*."[22] It is remarkable that enlightenment thinking by a native was first produced in the context of women's emancipation, given the highly patriarchal nature of the society at the time.

The novelty of Kartini's ideas went beyond a desire for modernity; rather, she dreamed for a "progressive and liberated world," an emancipation from both colonialism and feudalism.[23] Kartini writes: "We wish to equal the Europeans in education and enlightenment, and the rights which we demand for ourselves, we must also to give to others."[24] For Kartini, this equality was to be achieved through enlightenment practices such as education and science. Anderson argues that the idea of awakening from slumber, much like the metaphor of light from darkness, expresses transposition, a process of transformation of consciousness from traditional to modern. What is not clear, as Anderson himself admits, is the "indigenous ancestry" of this modern awakening.[25] Instead of seeing this concept as indigenous versus the West or something foreign, it is more accurate to think of the natives' lives as global; they had been impacted by international influences for over the past century, and this broader view had accelerated in the last decades of the nineteenth century.

The waves of enlightenment spirit were also felt in other parts of Southeast Asia, notably the Philippines. Much like the new *prijaji*, the Ilustrados (lit.

enlightened ones) refer to the educated Filipino class in late 1800s when the Philippines was under the Spanish rule. These middle-class Filipinos were educated in Spain and were exposed to liberal and nationalist ideals, and they inspired a revolutionary movement against colonialism. Among them was Jose Rizal, who had read *Max Havelaar* and engaged with Dekker's ideas.[26] Reynaldo Clemeña Ileto notes that an anonymous author in *La independencia* chronicles, in Tagalog, his passage from darkness to light. Echoing Kartini's expression, "darkness means condition of ignorance and death. His loneliness sets him on the road to enlightenment."[27] Likewise, the sailor Djamaloedin Tamin in chapter 7 recounts how the dangerous, lonely, and repressive conditions he faced as he mobilized the *pergerakan merah* clandestinely in Southeast Asian water only ignited his spirit for enlightenment. Pankaj Mishra documents a group of modern Asian intellectuals, including Jamal al-Din al-Afghani, Liang Qichao, Ho Chi Minh, Sun Yat-sen, Rabindranath Tagore, Ali Shariati, Sayyid Qutb, and Mahatma Gandhi, who wrote about the ideas of an emancipated Asia, national freedom, and racial dignity. This new spirit of intellectual awakening among the colonized reflected a larger trend across Asia. People from Egypt, Turkey, Persia, China, Indonesia, Vietnam, and Burma were awakened and rejoiced at the possibility of the rise of Asian power.[28] Even the African American leader W. E. B. Du Bois recognized this awakening as a worldwide outburst of "colored pride," as inspired by enlightenment emancipatory discourses.[29]

Kartini reflected the era of the Ethical Policy, when the welfare of the natives was addressed through the availability of education, the right to assembly and association, and access to modern media. Kartini's aspiration and references to enlightenment inspired early native modern organizations like Taman Siswa and Boedi Oetomo, which were both associated as the beginning of National Awakening, and native leaders. Among these figures was R. M. Tirtoadhisoerjo (1880–1918). Like Kartini, he was also of a *prijaji* background born to a noble family, but as a man he was able to receive Dutch education from a young age. Using journalism as his tool to resist colonialism, Tirto founded and ran the first newspapers in Malay that targeted educated native Indonesians, including *Poetri Hindia* (*Princes of the Indies*), to support the movement for women's emancipation. Tirto's activism went beyond journalism. In 1909, he founded Sarekat Dagang Islam (Islamic Trade Association) to advance the economic interests of the natives, which later became Sarekat Islam (Islamic Union, SI), one of the main engines of the *pergerakan merah*.[30]

These thinkers, from Dekker to Kartini to Tirto, helped to proliferate enlightenment thought in the Indies.[31] Their emancipatory ideas, in obscured translated forms, became part of the daily language and the aspirations of Indonesian natives after the turn of the twentieth century. From there, the *pergerakan merah* adopted enlightenment aspirations as they took upon novel ideas of communism to organize the first popular and radical anticolonial movement in the Indies. However, through the *pergerakan merah*, these ideas were no longer produced by the elite groups of *prijaji*, peranakan Chinese, Eurasians, or Dutch but rather by the *kromo* class, who produced their own literature and newspapers. In the next section, we explore how red enlightenment is manifested in verbal and symbolic speeches, notably in its lexicon, political slogans, visual culture, and literature.[32]

Red Enlightenment's Lexicon

Communist textual traditions in the Indies reflected a vibrant new anticolonial energy and enthusiasm, which was evident in the heterogeneous textual spaces of revolutionary newspapers. In addition to news on important events, article-length analysis, political slogans, and cartoons, a vernacular newspaper at the time would also publish literary works such as poems and short stories. Within these different textual genres, the red enlightenment was carried forward through the contentious lexicon of struggle and resistance.

Red enlightenment is a project produced for and by the *kromo* through creative activities of repurposing and cultural experimentation involving, among others, translation or transliteration, borrowing, repurposing, and adaptation of existing communicative tools and strategies for the purpose of mobilizing anticapitalist and anticolonial resistance. *Kromo* is a uniting identity that defined the movement. It is a Javanese term, often contrasted with *prijaji*, which, as previously discussed, began as a term for the Javanese aristocratic class, which had been recruited into Dutch service, holding key positions, such as *bupati* (regional heads), within the colonial hierarchy. Access to these positions at first was not granted based on education, but rather on descent. As Western education expanded to relatives of the aristocrats, the term *prijaji* extended to educated middle-class Javanese. *Kromo*, in contrast, referred to a loose sense of lower class and usually provoked the imagination of the larger population of landless peasants with no rank or status.[33] Connection between the *prijaji* class and the *kromo* class was at first highly mediated through a bureaucratic structure; however, with the rise of the new *prijaji* class, this gap shrank, and some educated *prijaji*, such as Darsono, joined the *pergerakan merah* movement.[34]

An overview of earlier vernacular newspapers such as *Doenia bergerak, Medan bergerak, Islam bergerak*, and *Medan Moeslimin* shows that "*boemi poetera*" (native), "*bangsa kita*" (our nation), and "*Muslimin*" (Muslims) were used to identify the targeted readers. The term *kromo* was likely used for the first time by Sarekat Islam and was later adopted by the *pergerakan merah* as a uniting identity of the targeted lower-class people. Even when the term proletariat replaced *kromo* (see chapter 5), meanings derived from *kromo* continued to shape the interpretation of the proletariat to include not just workers but also *orang ketjil* (lit. small, meaning poor people) and commoners (versus the elites). This is a unique product of a particular imperial history. It is also an unusual concept because it includes otherwise divergent groups: peasants, sailors, authors, day laborers, women, and children. The choice of *kromo* and proletariat, in lieu of *boemi poetera*, also shows how, at the time when nationalism had not taken a firm root in the *pergerakan*, the revolutionaries were able to imagine themselves as a part of a universal community of international workers.

The *pergerakan merah* produced a vocabulary of emancipation that was derived from enlightenment discourses and communism. While the previous native press had laid out the foundation of the vernacular press as a means for voicing criticism and protest against colonialism, the communist organ *Sinar Hindia* (later *Api*, henceforth *SH/Api*) furthered this legacy by creating a means for organizing collective actions and by maintaining a collective identity of the movement around communist language, concepts, and concerns.[35] The vocabulary of the *pergerakan* that had circulated for the past decade was repurposed to express communist ideas. Many of the phrases originated from Javanese culture, such as "*sembah djongkok*" (paying homage by squatting), "*sama rata sama rasa*" (equal distributions, equal feelings), and "*kaoem kromo*" (the *kromo*) were used as the daily language of the newspaper.[36] This vocabulary coexisted with widely known communist terms like "*kaoem Bolsyewik*" (the Bolsheviks), "*kaoem kapital*" (the capitalists), "*kaoem proletar*" (the proletariats), "*reactie*" (reactionary), "anarchist," "*geest*" (spirit), "*historisch materialisme*" (historical materialism), "*nationalisme*" (nationalism), and "*internationalisme*" (internationalism).

Likewise, the vocabulary of enlightenment that frequented Kartini's letters also figured into communist writings. Revolutionary leaders encouraged their members to mobilize the movement through a civil and harmonious way using intelligence and science. This is especially evident in *woro* Djoeinah's writings and speeches in *SH/Api*. Djoeinah was the first female editor of *SH/Api*, and the wife of a railway worker, Prapto, who was a leader in a railway workers union called the Union for Railway and Tram Workers (VSTP, see chapter 1). In her writings she would invite women from the educated class to join the *kromo* movement. In one of her writings, she states: "who will guide our wives [women]? A wife [woman] who loves us and has intelligence not only in her brain but also in her heart." In another article, she says "It is important that people can easily understand this request that knowledge [intelligence] is an important weapon to pursue in today's age."[37] The references to the enlightenment in the idea of "knowledge" and "intelligence" are clear.

In her writing and speeches, Djoeinah always emphasized that the movement should mobilize (*bergerak*) through a harmonious and civil way (*keroekoenan*) by upholding humanity (*kemanoesiaan*). The word *kemanoesiaan* peppered her writing and emphasized that the goal of the movement was for a universal emancipation of all workers, not just of the *kromo* in Java. She used *keroekoenan* in two different contexts. The first is to encourage that the people mobilized the movement in a harmonious way in solidarity.

> "*kaoem perampoean haroes djoega kita didik roekoen soepaja bisa koempoel*"[38] [women must be taught to gather in a harmonious way]

> "*merdika kita sebab orang sekarang bisa berkoempoel jang mana seakan akan melihatkan keroekoenan*"[39] [we (women) are free because we now can organize to show our comradeship/solidarity]

The second use of *keroekoenan* means "to organize" (verb) and "organization" (noun), which is a language that was borrowed from communism.

> "*Adjoekanlah kaoem istri dalam keroekoenan*"[40] [support women (to join) organization]

> "*kita telah mendapat sendjata oentoek melawan mereka, ja'ni sendjata 'keroekoenan dalam golongan sekerdja'*"[41] [we have found a weapon to fight them [capitalists], which is "organization (comradeship) of our class"]

Woro Djoeinah's verbiage explains that the fight against capitalism and colonialism must be done through the means that are, to use the period's word, "*berkesopanan*" (civil). Political and social organization was understood by Djoeinah as a civil and harmonious gathering based on comradeship. This explains why instead of arming themselves with weapons and using guerilla tactics, this movement produced and utilized revolutionary communication: *openbare vergaderingen* (public meetings, OVs), newspapers, pamphlets, books, schools, and global network of messengers. Through writing, reading, and debates, they produced knowledge that is inclusive, egalitarian, and global in their imagination of solidarity.

Teaching communist terminology and theory became an important part of the revolutionary press. In 1923 the Union for Railway and Tram Workers' (VSTP) head of printing, Partondo, was the first to translate Karl Marx's *Communist Manifesto* into Indonesian and published it serially in the *Soeara Ra'jat* newspaper.[42] Two years later, Subakat, under his alias Axan Zain, translated another version; both versions were printed at the VSTP's printing press facility.[43] Theoretical articles that were devoted to the explication of important communist terminology also commonly appeared in articles titled "What is Bolshevism?," "Revolutionary or evolutionary," and "Socialism."[44] In a new special Sunday edition (Sunday was usually an off day for *Api*), on November 22, 1925, *Api* published a translation of "Socialisme: Dari angan-angan hingga jadi pengetahuan (wetenschap)" (Socialism: From Dreams to Knowledge) by a Marxist and Comintern leader named Karl Radek. This was published along with articles "Jalan yang harus dilalui untuk membikin partai2 [partai-partai] jadi Bolsjewistisch" (The Route to Establishing Bolshevik Parties)," "Lantai komunisme" (The Floor of Communism), and "Marx and Engels" (Marx and Engels). On a special Sunday edition on November 22, 1925, the newspaper also published a discussion comparing feudal culture and communism called "Adat istiadat Minangkabau dan komunisme" (The Customs of Minangkabau and Communism) by a writer who worked under the alias Morgen Star.[45] These articles show the breadth of *SH/Api*'s publication, which included the promotion of the Comintern agenda.

Translation and transliteration were common tools in the movement.[46] They demonstrate active and creative processes involved in the movement that localized and appropriated foreign terms while inserting local terms into the communist repertoire. James Siegel captures this exciting moment of cultural experimentation, saying that

> [t]ranslation is not imitation, but it is not entirely different from it either; one repeats what the other said, this time, however, using one's own code. This assumes a code that was already formed, but, as we have seen, in the case of Melayu, one is uncertain.[47]

Translators actively put knowledge into the process and decision making that was involved in translation and transformed Melayu language by adding new terms like "*blenggandering-blenggandering*" (adopted from the Dutch word *vergadering* [meeting]).[48] Adrian Vickers notes that the aesthetic of imitation originated in ancient Java, but was also common in ancient West and Asian traditions. Speaking about Balinese paintings, he adds that artists "copy the works of others, but they produce highly original interpretations, both in terms of composition and style. . . . Artists strive to preserve the cannon of 'stylistic repertoire' while adding their own creative elements."[49]

SH/Api's rubrics also demonstrated its concerns about the importance of organizing collective actions. Just like other papers, a large segment of *SH/Api*'s space was dedicated to education, debates, opinion, analysis, and propaganda in rubrics like "*telegram*"—emergency news sent via telegram, "*warta-warta penting*"—important local, regional, international reporting, "feuilletons"—literary works, and *"surat kirim"*—letters to the editor. But, in 1923, a rubric called "Pergerakan" (movement) appeared replacing its old name "Kabar S.I." (SI news), which collected reports on *vergaderingen* (meetings) both public (*openbare*) and members-only (*leden*) held by the Red Sarekat Islam, Sarekat Rajat (People's Union, SR), and the Communist Party of Indonesia (PKI). This name change showed how public and members-only meetings were no longer seen as merely party activities, but rather as engines of a larger movement.

The style and tone of the written language of *SH/Api* is often propagandistic, agitating, and fierce yet dialogical, resembling the language used in strikes and rallies. The dialogical tone demonstrates the paper's purpose to target and address poor and illiterate readers; the editors knew that it would often be read aloud to an audience. Sometimes an article was followed with an editorial comment, such as this footnote from an article titled "Do We Want to Reconcile?"[50]:

> 7. Why with the misunderstanding, prijantoen [Javanese: noble man]! *Semaoen* said: "We don't have time to quarrel and be fussy" because we need to help workers to fight against the employers. If Marco likes, he can hold a vergadering [Dutch: meeting]. When the preparation is klaar [Dutch: done/ready], then we will come to debate. . . . Don't turn our words around like Abdoelmoeis, brother *Tjitro!*

Name calling and labeling, as well as the use of exclamation marks, characterized the dialogical nature of the language. But this oral-to-written style was also built through code mixing. While Low Malay was the dominant language, mixing it with Javanese and Dutch was common. Low Malay had been a lingua franca of the Malay Archipelago for centuries. It was first used by people of various races in the cities and commercial ports for commerce and trade but later developed into its own marker, differentiating indigenous people from foreigners.[51]

Apart from its dialogical nature, *SH/Api*'s language is also propagandistic and agitating. For example, this excerpt from an article explained what "capitalist" means:

> The capitalists are a group of people who live off of the labor of workers and peasants. In *Api* we have explained many times that what makes the capitalists powerful is *meerwaarde* [Dutch: value] (the surplus value coming from the labor of workers that are not returned to them).
>
> The trickery and deceit of the capitalists is very subtle that the people who are tricked and squeezed are not aware of it. They only think that their poverty and their bad destiny is because of God's will.

>   Workers and peasants who used to live and work alone are now gath-
>   ered in factories, trains, ships, and mining industries.
>   So, it is our enemies who *educate the heroes from our class*.[52]

This agitating language differed from the proper language that was promoted by
the government. In an article "Kekuasaan surat kabar" ("The Power of Newspapers") republished from the *Neratja* newspaper in *Sinar Hindia* on April 5, 1920,
the governor-general Alexander Idenburg called for the press to use "polite, and
sweet, language that does not offend anybody. This is called goede toon [good
manner].[53] If the press uses harsh language and create offense, however, this is
called slechte toon [bad manner]." This use of agitating language, so discouraged by the colonial authorities, served to further the movement's cause. *SH/
Api*'s "*slechte toon*" is reflected best in the newspaper's slogans, cartoons, and
feuilletons.[54]

Political Slogans

Persuasive oral expressions to evoke certain emotions and actions had been used
for centuries, usually in time of war and in religious texts and royal literature. In
the roarin' twenties, these expressions moved away from elitist and sacred texts,
and their production and consumption reached ordinary people. The *pergerakan
merah* adopted this technique through its publication of political slogans in their
own newspapers.

Just as the oral tradition made its way into the style of *SH/Api*'s written language, political slogans also crept into the prose.[55] Though often used in strikes
and rallies, *SH/Api* used political slogans throughout its pages, occupying empty
spaces in the margins of the newspaper to remind its readers the importance of
SH/Api and red enlightenment.

>   "How loud the voice of SINAR, to defend the Proletariat"
>   "The sun brightens the earth, SINAR brightens the workers' mind"
>   "Rice is the medicine for a hungry stomach, SINAR is the medicine for
>   a lost mind"
>   "SINAR demands equality between men"
>   "SINAR is the workers' mind"
>   "SINAR fights against capitalism and its slaves"
>   "SINAR is fierce, but also sweet"
>   "Being loyal to SINAR means defending humanity"

Notice how the words allude to enlightenment vocabulary of light and bright,
medicine, voice, equality, the mind, and humanity. During the time of strikes,

these slogans were used to encourage the spirits of the workers in a simple, agitating, and fierce tone. For example, when the VSTP strike launched on May 8, 1923, the following slogans appeared:

> "When on a strike, don't use weapons, just stay at home. Truly your employer will have a flat stomach."

> "The communists are always ready to pay back the devils' [capitalists'] exploitation! Beware."

> "Don't accept sweet words from the capitalists; they are all poisons! Believe in your [workers'] own strength."

> "Don't rely on fate. That is an old idea. 'Fate' only comes from reactionary's mouth."

The last slogan is representative of *SH/Api*'s attitude to the anticommunist group's use of Islam. Anticommunist propaganda often discredited communism for being non-Islamic and held that the only true way of life was by surrendering to the "fate" already written by God. But in the last slogan, the communist newspaper *SH/Api* argued that it was important to believe in one's strength and to not easily give in to "fate."

Political slogans were among the most important media that the central leadership of the communist party in Russia paid attention to. In reports from PKI leaders to the Comintern as well as meetings between them, slogans were carefully discussed and analyzed. Semaoen's report to the British Secretariat meeting on March 8, 1927, includes his observation that

> many of the slogans were not clear. . . . The masses did not understand them. This must be corrected. The slogans must be such that the masses will understand them.

He continues by describing some of the slogans used as "slogan for independence" and discussing how they avoided making the mistake of "using a slogan for dictatorship."[56] In a meeting with British Secretariat held on June 23 of the same year, Semaoen criticizes the communist revolts of 1926 to 1927 as lacking political preparation: "no slogans, no strikes, lack of contacts between the larger centers and the villages."[57] In other reports, the existing parties in the Indies were described by their different slogans—SI with its "Riches and wealth of the Islams for the Islam" and Boedi Oetomo with its "Java for the Javanese."[58] During the conference between Alimin, Semaoen, and the Comintern leaders, Semaoen suggested that the slogan "Down with the Dutch government" needed to be followed with "The United Kingdom of Indonesia," demonstrating how the creation of slogans involved serious discussions among top party leaders.[59] These slogans from top leadership contrast from those published in *SH/Api*. Despite the leaders' careful planning and crafting, *SH/Api*'s slogans show nuances in both content and expression and reflect concerns that ordinary rank-and-file members faced as they mobilized the *pergerakan merah*. These concerns went beyond the struggle against colonialism and capitalism and included a thirst for intellectual freedom, a search for modern education, a desire for equality, a defense for humanity, and a fierce resistance against capitalism and its religious and royal allies.

Cartoons: Symbolism and Imagery

As in other parts of the world where visual imagery was used to convey notions of resistance, *SH/Api* used cartoons to heighten the agitating effects of its verbal content.[60] In a meeting with the Comintern in Russia, the communist leader Darsono said, "we work through organizations, not only political organizations but also artistic organizations and various village units," showing how the use of visual propaganda was taken very seriously among international leaders.[61] It is not clear who drew these cartoons, as it was common for correspondents or journalists to conceal their names for fear of government persecution. The newspaper *Pemimpin* edited by Mas Marco, for example, was confiscated in 1921 due to a cartoon on its front page that was considered to be an insult to the government.[62] This was the first offense committed by a revolutionary newspaper that involved the use of an image. Because persecutions were rampant, writers and cartoonists commonly used pseudonyms such as "Moeda merah" ("Young [and] Red"), "Anarchist," "Si kromo boeroeh" ("The Lower-Class Labor"), "Sama rata" ("Equality"), and "Tjamboek" ("The Whip").[63] Generally, however, cartoons found in newspapers, movie advertisements, Komedi Stamboel, and comic, at this time were likely produced by the same artists working in the illustration business.[64]

The use of symbolic speech through visual culture has a long tradition in Indonesia going back to temple relief and *wayang* characters from a millennium ago. As John Lent documents, "Wayang kulit, as well as wayang bēbēr (scrolled picture stories) and lontar (manuscripts on palm leaves), all are known for their early telling of stories with pictures." Ancient *wayang kulit* (leather puppets) originated in the eleventh century carried humor, satire, and storytelling.[65] Anderson convincingly argues, however, that modern cartoons differ greatly from their predecessors due to their production using sophisticated printing technology and modern techniques.[66] Of a more recent past, Lent notes that "Jose Rizal is thought to be the first Filipino cartoonist and Ho Chi Minh the pioneering

Vietnamese political cartoonist."[67] Scholars also observe that cartoons in pre-independent Indonesia, like those in British Malaya, did not emerge until the 1930s.[68] The printing techniques used by Jose Rizal, Ho Chi Minh, and preindependent Malay cartoonists are unclear, as studies on the materiality of these early cartoons are largely absent. The analysis of political cartoons published during the *pergerakan merah* below shows that pre-independent Indonesian cartoons developed in the 1920s and that revolutionary printers were able to produce cartoons creatively in meager conditions to mobilize red enlightenment discourses in symbolic speeches.

The cartoons conveyed ideology and deconstructed myths, and both goals dovetailed with *SH/Api*'s verbal messages of red enlightenment. The headline "Capitalism or Communism That Leads to Sufferance" was accompanied by two cartoons.[69] The text under the drawing reads: "if workers are united and revolutionary, not only could they have a good luck, but they could also kick out capitalism like a water buffalo that headbutts a tiger and a wild boar. Therefore, workers, unite!" The cartoons are usually accompanied with poetic texts, such as proverbs, *pantuns,* poems, and rhymed sentences for their captions.[70] The text "workers, unite!" refers directly to Marx's famous line in the *Communist Manifesto*. The water buffalo, commonly used to work in paddy rice fields to plough the land, is used here as a metaphor for workers. It is hard working, strong, and powerful, and it is physically large and adorned with strong horns, making it unbeatable. The tiger represents the idea of a predator and an outsider while the wild boar represents an invader, both suitable metaphors for colonial capitalists who invaded and ruthlessly exploited the workers in the Indies. Within the context of colonial capitalism, the capital-owning class had primarily European backgrounds.[71] Water buffalo also figure into "beautiful Indies" paintings, alongside volcanos, rice paddies, and palm trees to evoke a picture of a timeless backwater where "everything is very beautiful and romantic, paradisical, everything is very pleasing, calm, and peaceful."[72] The cartoon represents a struggle that disrupts the harmony of the Javanese countryside.

In another cartoon, a monkey, symbolizing the capitalist class, is shown urinating on a water buffalo. The text says, "if workers are too patient, the capitalists will become insolent just like a monkey that urinates on the head of a buffalo." This image invites workers to be active and persistent in fighting against the capitalists. In preindependent Malay cartoons of 1930s, monkeys were often used to symbolize "Malays who are aping Western attitudes." The monkey here likely represents natives who supported colonial capitalism and acted like capitalists.[73]

FIGURE 1. Cartoon "Tiger and water buffalo"

FIGURE 2. Cartoon "Monkey and water buffalo"

The use of animal imagery was common in revolutionary newspapers. Another cartoon found in *Sinar Hindia*, titled "Crocodiles Created a *Vergadering*," depicts

> how crocodiles have lost their minds to influence their people that it created a chaos in their own heads.
>
> On the sly, after he opened the congress in Garoet where he created various *voordracht* [Dutch: speeches], he then went on to Bandoeng to open a second *vergadering*.
>
> Those who can guess who the crocodile is can receive *Sinar* [SH newspaper] for a week for free.[74]

Satire was not intended to break news or provide new information on an event. Rather, its effectiveness relied on intertextuality, generating relationships between separate works. Cartoons are frequently used to respond to news and events.[75] This cartoon recalled the report of top leaders of White SI, including Tjokroaminoto, who held a secret meeting in Garut. Crocodiles symbolized traitors—in this case the White SI leaders, who expelled the communist-leaning members of SI party. In the picture, two monkeys, one of which likely refers to Tjokroaminoto,

Boeaja   memboeat   vergadering.

FIGURE 3. Cartoon "Crocodiles created a *vergadering*"

a White SI leader who aped the capitalists, foreground several crocodiles that are having a meeting.

This symbolism reveals the place of animals in the broader history of the Malay world. Tigers were important in the imagery of the West, especially in its characterization of its other, the Orient, at least until after the Second World War when Western countries lost their colonies and tigers "ended up as bedtime stories" for children and a mere "symbol of the Empire's faded splendor" for adults.[76] However, for Malay speakers, tigers have long been important characters in myths, legends, fairy tales, and fables and can be found in paintings, carvings, and sculptures, as well as in Javanese and Balinese shadow performances called *wayang*.[77]

The characterization of tigers as predatory capitalists in communist cartoons borrowed from common ideas that had circulated in the Indies for centuries. Tigers were known as "man-eaters" that killed and ate wild boar, along with a wide range of other game.[78] Hunting and trapping of tigers to rid the environment of dangerous animals could be traced as early as the 1600s, when the sultan of Banten gained the title of "a tiger hunter." Hunting tigers would later be joined by European inhabitants and supported further by the expansion of the Dutch colonial state between 1800 and 1870.[79]

This symbolism was rooted in nineteenth century rituals that involved fights between tigers and buffalo. Often performed as entertainment for European visitors, Thomas Stamford Raffles reported that "[i]n these entertainments the Javans are accustomed to compare the buffalo to the Javan and the tiger to the European." While fierce and dangerous, the Javanese believed that the tiger would be worn out by the buffalo with its "formidable staying power."[80] Peter Boomgaard notes that "one can also imagine that the ceremony had yet another meaning for the peasantry. They may have seen it as an encounter between agriculture (the buffalo, plow animal to many Javanese) and "wild" nature (the tiger)."[81] The first Indonesian painter, Raden Saleh (1807–80), is famous for his use of animal imagery to express social criticism, and earned twenty years of Western-style painting education in Europe before appointed to be a royal painter in Java.[82] One of his famous paintings depicts a fight between a buffalo and a tiger, which gives us a broader context into how the communist cartoons were likely inspired by and repurposed existing circulating imagery by Saleh.

The appropriation of visual tradition by the revolutionaries is also evident in the use of *wayang* symbolism. Here, the Javanese character depicts a member of Volksraad (People's Council) from Boedi Oetomo, a *prijaji* organization. The Javanese clothing worn by this character alludes to a *punakawan*. In *wayang* drama, the *punakawan* are usually servants, followers, and dependents. They follow where their masters lead them, and are permanently positioned as their subordinates.[83] However, the *punakawan* are usually shirtless. Iconography can also establish ranks. The Javanese character here was adorned with turban-like headdresses and wore a coat-like garment over their sarongs, which alluded to priests and other characters of spiritual power.[84] This mix of references implies that the character in this cartoon, despite being in a place of power, was a servant of the colonial governments. It evokes the circulating sentiment that Volksraad's power is limited even though they supposedly represented the people. The text saying "Dear Mr. Chairman! It is true that I am a Javanese, but I know customs and manners" underlines this understanding. It satirizes Javanese who hid behind traditional customs and manners and were reluctant to use their power against the colonial government because of the benefits they gained from their position.

Some of these cartoons show an act of creative repurposing. Under the first cartoon "Tiger and Water Buffalo," the text states "Cliché dari Sin Po" (Cliché

FIGURE 4. Cartoon "B. O. in Volksraad"

from Sin Po). Sin Po is a radical peranakan Chinese Sino-Malay newspaper published in Batavia, which showed support for the Indonesian anticolonial movement during the 1920s.[85] With a low budget and technological constraints, *SH/ Api* borrowed these illustrations from other newspapers which were sympathetic to its cause, including Sin Po, and reappropriated the images to illustrate communistic rhetoric and propaganda. This form demonstrates the wider context of typography of the period. Instead of using woodblock—a lower budget production—illustrations at this time likely utilized cliché or stereotype, a solid plate of type metal cast from a *flong* (a temporary negative mold) that is made of a form of set type. Because of the skills and technology required to conduct the process of stereotyping, it is possible that identical clichés were produced and sold—or, in the case of *SH/Api*, borrowed—from a number of printing houses in the Indies and abroad. This technology allowed these images to be produced and circulated across varied contexts and textual environments.

A closer analysis of the cartoons helps us understand the material context of their production. The first two cartoons, "Tiger and Water Buffalo" and "Monkey and Water Buffalo" both carry a signature of the artists (see the lower right corner), indicating that they could be recognized as high art. The printed lines show that they involve complex cut. On the other hand, crocodile and *wayang* cartoons are less complex. The lines surrounding the *wayang* cartoon on the top and left sides seem unintentional, demonstrating that these were not intended for fine art printing. Working from a place of scarcity—and often illegally, as indicated by Mas Marco's arrest, these cartoons were produced quickly to carry didactic political messages.[86] (Other communist cartoons are available in the appendix.)

Communist cartoons demonstrate a shift in both the production and consumption of visual culture from elites, like Saleh and the royal communities, to commoners who often lacked the same expertise and knowledge. This does not mean that one replaced the other, but rather that the source of power is no longer concentrated within the few. With access to printing technology and techniques, the *kromo* could produce their own visual culture.

Feuilletons: The Birth of Socialist Literature

Besides lexicon, political slogans, and cartoons, the red enlightenment project was also voiced in poems, novels, and short stories—literary styles which first emerged and were popularized through the vernacular press. If the birth of native vernacular press was highly influenced by peranakan Chinese press, so too was the development of native literary genres. Between 1870 and 1880, Indonesian–Chinese literature appeared in peranakan Chinese newspapers and soon became popular among peranakan Chinese readers, who would send and contribute poems and short stories to the newspapers. Fictional stories would usually be published serially in the feuilletons (the French word for the literary and art section of a newspaper) and only later be bound and sold in booklets or brochures to expand sales and readership.[87]

Against the backdrop of the popularity of literary fictions in vernacular newspapers, revolutionary newspapers such as *SH/Api* utilized literary genres to popularize emancipatory ideas. These radical fictions were labeled "*literatuur socialistisch*" by Semaoen, but the government called it "*bacaan liar*" (wild readings).[88] According to Semaoen, "socialism is knowledge that regulates social life in that humans do not exploit one another."[89] Socialist literature educated the *kromo* class about the existence of colonial capitalist exploitation and motivated them to organize and overcome that exploitation. Some famous works published as a series in *SH/Api* and then later as a novel were Semaoen's *Hikajat Kadiroen* (The Tale of Kadiroen)[90] and Soemantri's *Rasa Mardika: Hikajat Soedjanmo* (A Sense of Independence: The Tale of Soedjanmo).[91] Written during Semaoen's time of seclusion in jail, *Hikajat Kadiroen* recounted a story of Kadiroen's adventure as a political organizer. It was framed around a quest to win Ardinah's heart. As a native young government official, Kadiroen was frustrated to see how, despite having already worked hard to help them, *kromo* people still lived in poverty. Realizing this, Kadiroen, like many other fictional characters in this genre, joined the PKI to help liberate his people from capitalist exploitation. In another story, the main character reflected this trend by summing it up as "*bekerja atau bergerak*" ("to work or to move"). This mirrored the real-world antagonism between anticommunist employers and communist workers.[92] *SH/Api* often reported about workers leaving their jobs because of their involvement in the movement, as urged by *Hikajat Kadiroen*.[93] This daring call provides a context in which, as H. M. J. Maier describes, "ordinary people began to turn away from their traditional leaders" looking for new leaders and a new foundation.[94] A new age means a communist age.[95] This story appears crude and didactic, and lacks the aesthetic of high literary culture; it was intended for mobilizing and came with its own aesthetic.[96]

A. Teeuw recognizes *literatuur socialistisch* as "a new literary genre" of its time, giving a new direction in the development of Indonesian literature. Calling Semaoen's *Hikajat Kadiroen* one of the first examples of this genre, he says that "although these novels primarily served political and ideological aims," they depict "stories about contemporary people, based on real contemporary, political, social, economic, and cultural conditions and situations."[97] In these stories, daily processes of anticolonial organizing were told within the interstices of romance and love stories. James Siegel argues that romance became an appealing metaphor for conceiving national independence because it was the way to represent the tension between feudal and colonial styles of hierarchy and provides a third way, "nationalist love." "[T]he morality of nationalist love upheld against familial authority not only reclaimed familial authority for the nation but made Europeans the locus of immorality."[98] Freedom to choose a marriage partner was apt; it was one of the most private and personal life experiences in which nationalist sentiments could be evoked against both feudalism and colonialism. Socialist literature, however, uses romance in a different way.

Although these works predated the debate among the postindependence Indonesians left regarding the application of socialist realism in Indonesian literature, these radical fictions promoted socialism and communism as a promising way to liberate workers and the *kromo*.[99] Often defined as, in Raymond Williams's words, "full, central, immediate human experience," literature enacts sensibility and desires differently than in other nonfiction genres.[100] To convey exploitation and inequality in colonial society in the immediate living experience of the main character, Kadiroen helped readers to emulate his real-life choices and attitudes. The use of fiction to promote communism was not intended for just a mere creation of a "revolutionary taste" but rather as a part of an emancipatory and revolutionary project to organize the working class.[101]

The efficacy of literary fiction to propagate communist ideas cannot be understated. One way to assess this is by looking at the subsequent feuilleton stories published in *SH/Api* in 1925. Four short stories were serially published in *Api* that year:

1. (Story 1) "Korbannja nafsoe birahi atau godanja pertjinta'an jang soetji" ("The Victim of Lust or the Seduction of Sacred Love") by Tjoa Moh Shan and Ong Kiong Giam in *Api,* January 5–February 12, 1925
2. (Story 2) "Rohmani: Kekejaman iboe tiri" ("Rohmani: The Cruelty of a Stepmother") by Djamboemerah (alias Red Rose Apple) in *Api,* February 14–March 19, 1925
3. (Story 3) "Siti Maryam, perempoean jalang yang berboedi" ("Siti Maryam, a Virtuous Bitch") by Djola Djali (an alias) in *Api,* March 20–May 12, 1925
4. (Story 4) "Aliran djaman atau seorang gadis yang sengsara" ("The Flow of Time or a Girl Who Suffers") by Tjempaka-Pasoeroean (alias a *tjempaka* flower from Pasoeroean) in *Api,* August 6–September 10, 1925

In 1925, *SH/Api* had become even more uncompromising and concentrated more in aiding the radical and popular *pergerakan merah* on the ground, so publishing four short stories with a focus on love was peculiar. Before that year,

SH/Api published more outwardly communist, albeit still romantic, stories like *Hikajat Kadiroen*.

It is important to situate the production of *literatuur socialistisch* within the larger context of Balai Poestaka publication. As the largest publishing company funded by the colonial government, Balai Poestaka set the taste and desire of popular readers, and most of the modern books were original modern Malay novels and translations of Western literary products. Of the translated books, the most popular ones were romantic adventure stories because "[o]ne can hardly call the [translated] books . . . basically different from Indonesian literature."[102] Love is a universal theme and, hence, universally easy to sell. However, the similarities between these mainstream fictions and those published by communist presses stop there.

Although the four short stories in 1925 did not center around the story of a communist party member like *Hikajat Kadiroen*, they still employed the movement's vocabulary and concepts as well as the theme of liberation, demonstrating how communist language no longer belonged to the leaders. Instead, it circulated among ordinary readers who contributed the stories to *SH/Api*. The authors were probably party leaders who wrote under pen names. Among the four stories, "Aliran jaman atau seorang gadis yang sengsara" was likely assumed to be written by a party leader, as evidenced by its setting in the radical city of Semarang and its inclusion of detailed scenes of *vergadering* and discussions of communist concepts. Because Semaoen was already abroad by 1925, it is possible that Tjempaka-Pasoeroean was Soemantri, who would have composed this story in jail, where he also wrote *Rasa Merdika: Hikajat Soedjanmo (A Sense of Independence: The Tale of Soedjanmo)* in 1924. However, the true authorship of the story has not been confirmed, and previous scholars have made the mistake of associating stories written by unknown authors with more famous or experienced writers. Anderson, for example, contends that Synthema was the pen name of Mas Marco, while in fact it was Soemantri's.[103] The story's epilogue, however, casts doubt on this notion. Tjempaka-Pasoeroean writes

> Actually we are not an expert in composition. However, because we want to learn to write, with difficulties due to low knowledge, we force ourselves to start writing bit by bit until now it becomes a feuilleton titled "Aliran Djaman" published in this *Api*.[104]

Pasoeroean's statement implies that the writer is not an expert in composition who autodidactically learned to write. This nonexpert claim is important; the writer created an identity of an ordinary person and distanced herself from the rank of "expert" editorial leadership in *SH/Api* newspaper. This distancing ruled out Djoeinah, the only female editor at SH. A further close reading of the text shows the writer's sensitivity and attention to women:

> Among the flowers is a young woman wearing a green kimono. . . . Her long hair is let loose reaching her heels.[105]

> Poor brother Semaoen! He thinks about the fate of many people that his body has to be separated from his wife and children.[106]

Regardless of their specific authorship, these stories reveal the existence of ordinary writers—even if they were rank-and-file leaders—outside of the short list of famous communist writers, for example, Semaoen, Soemantri, and *woro* Djoeinah, who were active during the period. It demonstrates the widespread and active involvement of ordinary writers, and how the movement had evolved into a more democratic and egalitarian entity. These "nonexpert" authors actively produced red enlightenment discourses of struggles and freedom. At least two of them did not share a background of the editorial board of *SH/Api*: One has a Chinese name and another, Pasoeroean, seems to be a woman.

These "nonexpert" authors choose women as the protagonists. Story 1 narrates how a young *peranakan* Chinese woman, Tjian Nio, lived with a widower uncle and cared for his four children. Story 2 centered on Rohmani, who lived with her father and his cruel new wife. Story 3 was about Siti Maryam, a beautiful Sundanese woman who fell into prostitution. Story 4 was about Soediati, the daughter of a police officer who forced her to marry an aging government official for money. In each of these stories, the female character faced a condition that was common at the time—an uncle almost raped Tjian, Rohmani and Soediati were about to be forced to marry a man for money by their respective parent, and Siti Maryam became a sex worker to survive. Eventually, by finding their sense of self as women within their oppression, these women were able to find refuge: Tjian, Rohmani, and Soediati ran away to other cities and found the loves of their lives, while Siti Maryam left sex work and lived with her boyfriend.

These accounts may seem overly sentimental and moralizing. But taken in the historical context of their production, their themes highlight the concerns of the moment: enforced domesticity, arranged marriages, and limited access to love marriages. That these women could run away from their homes to liberate themselves and to determine their own affective futures was a new phenomenon inspired by the language of the movement.

Much of the revolutionary literature was written as love stories, and the concept of love marriages was used as a vehicle for nationalist and revolutionary zeal.[107] But these stories centered on men, while the four *SH/Api* short stories focused on women. Sylvia Tiwon states that postindependence New Order narratives, selectively built on prior narrative traditions, positioned women in stories, not as narrators but as "vessels" of pathos.[108] Their suffering is the site through which the reader comes to feel. In the *SH/Api* stories, these women's suffering, and ultimately happy endings, were meant to generate feelings of anticolonial passion. More importantly, they show both the centrality of women characters and their potential to be recognized as full heroes. Later nationalist narratives, most obviously in the figure of Kartini, positioned women as contributing to national liberation through sacrifice and even death. By contrast, the women in communist fiction not only survived; they thrived.

Red enlightenment concepts and vocabulary of liberation and new ways of thinking and living were applied in the language as well as the social commentaries embedded in these love stories. A selection of excerpts illustrates this:

> money marriage (story 1)

> Our romance is degenerated because of wealth (story 1)

> Don't believe in superstition [because] it doesn't rely on truth (story 1)

> It is an old-fashioned understanding to say that women had to stay at home (story 2)

> Those [Dutch] women felt embarrassed and lowly . . . to sit at the same level as their maids [who had to sit on the floor] (story 2)

> Those poets who think in an old-fashioned way assuming women as just objects for entertainment or display (story 3)

> people these days still feel weird to see a love union between two people with different races (story 3)

> Soediati sit on the floor facing her parents who are sitting on the chairs. Actually Soediati does not like to follow this old-fashioned rule. . . . In this day and age *sembah djongkok* [paying homage by squatting] is no longer appropriate. (story 4)

Fictional stories published by Balai Poestaka around the same time, most notably Abdoel Moeis's *Salah Asuhan* (*Wrong Upbringing*) and Marah Rusli's *Siti Nurbaja* (*Siti Nurbaja*), also focus on the clash between the modern and the traditional. Through *Siti Nurbaja*, Rusli—a Minangkabau intellectual professing himself as an adherent of modernity—makes "a plea for free choice, by young people, of their marriage partner, for monogamy, and for the responsibility of fathers for their own children."[109] Similarly, Moeis's *Salah Asuhan* also depicts an interracial romantic relationship between the *bumiputera* (native) Hanafi and the

Eurasian Corrie to show this clash between modernity and tradition—including a criticism against *sembah djongkok*. But in the end they are both soundly punished for violating traditional norms.[110] While they support a modern way of life, none of these stories was intended to raise an understanding of universal emancipation and solidarity, let alone class consciousness. In fact, Balai Poestaka intended to counter communism by using "tactic of subtle manipulation" in popular entertainment novels.[111] According to Jedamski,

> The first example [of these novels] dealt with a regent's daughter who rejected being married according to her social standing, but who married a carpenter instead—and finally led a life full of joy and happiness. Ziesel [a Balai Poestaka officer] admitted that this adaptation of an originally Western text was well done and smoothly written. As an appropriate counterpart Ziesel imagined the story of a regent's daughter who takes all efforts to explain to a workman, who has fallen in love with her, that there is no happiness to be found outside one's own social class.[112]

While the story shows how the regent daughter was able to marry her chosen carpenter man, it subtly advises that he would also be happy to remain within his social class. In these stories, class differences are not a problem if one could make their own life choices, romantic and otherwise. This counters the communist narratives that capitalism creates two opposing classes—the capitalists and the workers or peasants—and how it is necessary to overcome these class differences and the conflicts emerged out of them.

Tirtoadhisoerjo's short stories also influenced and shaped *literatuur socialistisch*. Around 1909, Tirto published a series, *Nji Permana*, in his newspaper *Medan prijaji* that told a story of Permana, a child of a village chief who was forced to marry a corrupt police officer by her feudal father. The soul-searching battle finds Permana at a stage where she must choose between staying at the village and creating a new life. "Going beyond his time," as Pramoedya Ananta Toer describes him, Tirto shows at the end that Permana chooses instead to ask for a divorce from her husband.[113] Tirto's other story, "Cerita nyai Ratna" ("The Story of *Nyai* Ratna") also depicts a thriving female character, Ratna, who appears to get away with all sorts of things, including murdering her husband. In the end, she survives without serious moral consequences.[114] Strong women characters might find their predecessors in Tirto's stories; however, the female protagonists in communist fiction are about more than showing how to win a battle against men.

By contrast, the thriving female characters in *literatuur socialistisch* fight the problem of double erasures of women in Indonesian literature.[115] In both revolutionary literature and romantic fiction, they were either completely absent or present only within the confines of existing power structures, either through the reproduction of their stereotypes or through the articulation of an identity that did not challenge the structures that limited them—colonial, nationalist, or otherwise. This includes later writings advocating for women's liberation such as in the founding president Sukarno's 1951 book, *Sarinah: Kewajiban wanita dalam perjuangan Republik Indonesia* (*Sarinah: The Responsibilities of Women in the National Revolution*). Despite promoting the emancipation of women, revolutionary femininity was still subsumed under the larger projects of nationalist and anti-imperialist agenda. Women and their struggles continued to be perceived as secondary, and their material realities of inequality remained unaddressed by the male leaders of Indonesia's national revolution. The discourse on women's emancipation was only evoked as a strategy to gain women's support for national unity.[116] *Literatuur socialistisch* of the 1920s was different. It represented women as successful heroes and gave them active roles. In these stories, women organized the *kromo* class in political resistance against capitalism and colonialism and affirmed that the new way of thinking only relied on truth, saw women as free-thinking beings, perceived people of different races as equal, and disapproved marriage based on money and old ways of paying respect. The fact that these commentaries were found in fictional stories with romantic themes, which shielded their focus on communism, shows how the new way of thinking was most relevant in mundane seemingly nonpolitical areas. By the 1920s, communist modes of thinking had become the language of the ordinary.

Cultures of Resistance

The roarin' twenties, characterized by cultural experimentation, creative adaptation, and mass production and rapid circulation of ideas and information helped birth the red enlightenment. Enlightenment aspirations of truth, pursuit of knowledge, and emancipation were adopted and mobilized in a courageous battle to express anticolonial and anticapitalist critique and resistance. This *kromo*-led movement translated communism within the existing framework of enlightenment. In the meantime, enlightenment became absorbed in the fierce and radical language of communism. Education, truth-seeking, literature, *keroekoenan* (organization/comradeship), and debates became the main weapons of class struggle and *kemanoesiaan* was the motivation to fight against exploitation and injustice rooted in colonialism, religious dogma, and monarchism.

The red enlightenment existed within a context in which access to the production and consumption of literature and arts had expanded from the elites and the middle class to the *kromo* and the commoners. At least two conditions allowed for the itinerary of emancipatory thoughts to travel via the van Hogendorps, Dekker, Kartini, and Tirto to the *kromo*. They are the expansion and access of modern printing technology and the arrival of the radical politics of communism. Taken together, the tradition of verbal and symbolic speeches reflected in the *pergerakan merah*'s lexicon, slogans, cartoons, and feuilletons became some of the main engines to create and normalize a new system of thoughts. The use of the new language of resistance and enlightenment in everyday mundane life made the new system of meaning an everyday exchange creating cultures of resistance. This too is an important characteristic of the *pergerakan merah*; ordinariness was important to both the technique and imagination of the anticolonial resistances. In ordinary cultures of resistance, women figured prominently in their agenda.

The communist verbal and symbolic speeches were mediated through revolutionary communication, a set of communicative practices and technology produced by the *pergerakan merah*. They include OVs, People's Schools, the revolutionary press, and sailors as messengers. How did the revolutionary communication embody the emancipatory projects of red enlightenment? This question will be tackled as these different forms of revolutionary communication are discussed in subsequent chapters. Before we explore them further, however, we will first delve into the geography of resistance to understand the composition of the *pergerakan merah* based on a set of rich and original historical geographic information system maps on the expansion of the movement.

GEOGRAPHY OF RESISTANCE

The prevalence of public meetings was key to the spread of red enlightenment, the growth of the anticolonial resistance, and *pergerakan merah*'s eventual demise after the 1926 to 1927 revolts in the Dutch East Indies. From 1920 to 1925, over 800 public meetings led by the *pergerakan merah* were held across the Indies archipelago in both private residences and public parks and were attended by up to 15,000 people. Abroad, Indonesian sailors working on Dutch shipping lines organized clandestine meetings aboard the ships and in the sailors' quarters in Holland, often including sailors from other countries. When most of the central leaders of the party had been exiled in 1923, public meetings continued to appear in urban and rural areas of Java and the Outer Islands.

A look at the communicative practices of this movement points to an oft-neglected understanding of how communicative practices reveals the ways in which space and power operate. Utilizing historical GIS (geographic information systems)—a research method that captures changes over time and space, this chapter conducts a spatial analysis to investigate the role of public meetings (*openbare vergaderingen* [public meetings], OVs)—the public sphere of the colonized in the Indies—in the popularization and radicalization of the *pergerakan merah*. It argues that resistance against colonialism was as much a spatial struggle as it was a political one. Social movements are often seen as harbingers of social change, and thus are always emergent and transient. As both a research method and a mode of representation, GIS mapping can enable us to see how resistance in the Indies was communicated in transient time and space.[1] The development of these maps acts as a counternarrative and facilitates resistance to the hegemony of authoritative maps created by the colonial state powers that we discuss in chapter 1.

Chapter 1 discussed how the implementation of modern transport networks created a condition of possibility for new forms of mobility, and, hence, a new form of organizing resistance in the Indies. This suggests that the mobility of people across remote regions during the movement was relatively accessible, and that transport and communication workers played a key role in the mobility of the red enlightenment's emancipatory ideas. The infrastructure of colonial capitalism facilitated the emergence, expansion, and active mobilization of the *pergerakan merah*. In other words, the infrastructure of colonial exploitation became an infrastructure of resistance.

This chapter furthers this exploration by studying mobility from the perspective of the political movement of people and ideas. It provides visual and statistical evidence for the expansive spread of the movement, the diversity of its participants, and new important timelines prior to the communist revolts of 1926 to 1927. An analysis of this set of historical GIS maps, based on the reports of communist OVs that were held across the Indies archipelago in the period 1920 to 1925, reveals the development of the *pergerakan merah* over space and time, the spatial and social characteristics of the movement as reflected in the gender and racial identities of the attendees, the speakers and the topics that were discussed, the evolution of the political parties, the sites of the meetings, and the pervasive surveillance over the daily activities of the movement by the governmental police. These maps illustrate that the movement was mobilized, for the first time, across widespread geographical areas in the Indies archipelago—it was not just centered in the island of Java—and that it was organized across different cultural borders and identity markers—nation, gender, class, and ethnicity. This widespread spatial and cultural scope enabled women, rank-and-file members, and people of diverse races and ethnicities to participate in the revolution by meeting in mundane spaces like movie theaters, private houses, and school buildings. This chapter also provides evidence for the first time that 1924 and 1925 were the peak years for the mobilization of the movement from below, and that it occurred when most of the main leaders were either in jail or exile. It offers a different perspective of the revolts and its aftermath.

The Development of the *Pergerakan Merah*

The foundations of native modern political organizations in the form of unions and political parties emerged in the period of what Ruth T. McVey calls "the Ethical policy of Enlightenment" at the turn of the twentieth century.[2] In this period, colonial government began to allow access to modern education, political

assembly and organization, and the press for native population. The first modern native political party, Boedi Oetomo, was founded in 1908 by the "new" or
"lesser" *prijaji* of Java. The lesser *prijaji* comprised of a group of indigenous people who were taught in the Ethical Policy's modern education system and were
making their careers in government services. The formation of the new lesser
prijaji became the backbone for the expansion of the Dutch administrative system (about 90 percent of government officials in the Dutch East Indies by 1931
were natives occupying positions in government offices, police institutions, and
courts).[3] Boedi Oetomo was a primarily Javanese *prijaji* organization that promoted the idea of emancipating Indonesians through Western education and was
predominantly attended by the *prijaji* class. Starting in 1908, new organizations
of educated elites proliferated. Many of them were based on ethnic identities: the
Eurasian (European and Asian, mixed blood) Indische Partij (Indies Party, 1911),
Jong Java (Young Java, 1915), Jong Sumatranen Bond (Young Sumatrans Union,
1917), Jong Ambon (Young Ambon, 1920), and so on. Still, these organizations
had not successfully attracted a mass following among the uneducated, lower-
class natives.

The foundation of two political parties, Sarekat Islam (Islamic Union, SI) and
Indische Sociaal-Democratische Vereeniging (Indies Social-democratic Association, ISDV) would change this and later would become the basis of the *pergerakan
merah* in the Indies. Tirtoadhisoerjo, having left government service to become
a journalist, founded the Sarekat Dagang Islamiyah (Islamic Commercial Union,
SDI) in Batavia in 1909 and in Buitezorg (Bogor) in 1910. This organization
was designed to support Indonesian traders, and the use of the term Islam in
the name of the organization was a uniting identity to reflect that the Indonesian trader members were Muslims while the Chinese and Dutch competitors
were not.[4] Branches in other regions were soon established. In Surakarta, Haji
Samanhudi led the SDI branch as a Javanese batik traders' cooperative. The Surabaya branch was founded by H. O. S. Tjokroaminoto, a charismatic figure with a
hostile attitude against the authorities who would soon take over the leadership
of SDI and change its name to SI. Under the leadership of Tjokroaminoto, the
Islamic and commercial origins of the organization were replaced with resent-
ful voices against the government, giving the organization a reputation as the
savior of the lower class and Tjokroaminoto as the *ratu adil* (the "just king"), a
messianic character in Javanese legend. From 1912, as the organization spread
throughout the villages, the membership of SI rapidly swelled, and many uneducated lower-class natives joined the organization.[5]

Amid SI's expansion, the ISDV was founded. It was established by a young Dutch
labor leader, Hendricus Josephus Franciscus Marie Sneevliet, who had arrived in
the Indies in 1913 in search of employment. At the time, educated Europeans
were in high demand in the Indies, and Sneevliet—who had left Holland due to a

disagreement with his fellow Dutch socialists—had no problem finding work as an editorial member of the *Soerabajaasch handelsblad,* a principal newspaper in East Java and the voice of the powerful sugar syndicate based in Surabaya. Soon after, he would move to Semarang to take a job in the Semarang Handelsvereniging (Commercial Association). As a product of the Ethical Policy's goal of raising the Indonesian standard of living, Semarang had become an expanding urban port city in Java, the center of European commerce and the base of radical activities in the Indies. It was also the headquarters of the Indonesian railroad workers' union, the VSTP. As a talented propagandist who had moved from Catholicism to socialism in his search for salvation, Sneevliet's firebrand personality was a good match for the VSTP. But helping the VSTP to publish its *De volharding (Persistence)* newspaper was not enough. On May 9, 1914, he initiated a gathering of sixty social democrats—mostly Europeans—to found the ISDV, whose goal was to propagate socialist principles, notably revolutionary anti-imperialism, in the Indies, which they believed would help defeat colonialism.

It was one thing to create an agenda to participate in the Indies politics; it was another to actually put it into practice. With most of the association's original members being European, how could their propaganda succeed? Since regulations did not prohibit people from joining more than one political organization, the ISDV would soon find followers through existing organizations. They did not achieve much success with the Eurasian-oriented socialist group Insulinde, but they did with the VSTP and SI. The VSTP was dominated by radical native railways activists, while SI had no comparable standing in terms of its popular following among lower class natives. Together with Sneevliet and A. Baars (the editor of ISDV's organ *Het vrije woord*), Semaoen—a fifteen-year-old talented VSTP leader—and other Indonesian members attended and addressed SI gatherings and kept close relations with its leaders. As a result, SI became even more radicalized. By the time that Sneevliet was exiled and forever banned from reentering the Indies in 1918,[6] the ISDV had already picked up a generation of Indonesian leaders who would continue its program and, in 1920, turn the party into Partai Komunis di Hindia (later Partai Komunis Indonesia, PKI), the first communist party in East and Southeast Asia.[7] Together with SI, the PKI and other affiliated trade unions would become the basis of the *pergerakan merah*—the first nationally and globally connected popular anticolonial movement in the Indies.[8]

Mapping the Movement

The geography of resistance represented through the GIS maps of OVs provides new visual evidence of the popularization and radicalization process of the *pergerakan merah*. It complements works by earlier writers such as Anton

Lucas's oral history of Peristiwa Tiga Daerah (the Three Regions Affair), which painstakingly collects local cases to map resistance movements in 1945.[9] The data set in this chapter allows us for the first time to see the *pergerakan merah* of 1920s in a broader picture and clarifies earlier assumptions. The data comes from 826 reports on OVs held across the Indies archipelago between 1920 and 1925 gathered from *Sinar Hindia/Api (SH/Api)*.[10] These reports include detailed information on locations, dates, speakers, attendance, organizations, meeting sites, topics, and government interventions. These details were first coded and then were processed in statistical analyses using the software SPSS Statistics.[11] The statistical analysis generated using these data on public meetings provides added information and a new perspective in looking at the dynamics of public meetings. It demonstrates, for example, the importance of the involvement of women and non-indigenous people within it from 1923 through 1925. Using SPSS, correlations were also made between locations and other variables such as speakers, number in attendance, women's attendance, and topics discussed, as well as the dates of the meetings and the kinds of government interventions. These were then visually turned into GIS maps. This set of maps describing the spatial changes of the movement vis-à-vis its social, political, mental, and cultural compositions shows what I call "the geography of resistance."

Although the names of villages are known, the exact locations of the public meetings so far have been grouped only under the residency level due to the absence of a primary map from the period. For this, I regrouped about 182 location names—each consisting of a number of villages, collected from the location information of the 865 meetings, into twenty-two residencies, using Robert Cribb's map of 1931 residency boundaries in Java island as a model to create the boundaries between the residencies.[12] Residency names in this chapter, hence, follow their original spellings. For the GIS maps in this chapter, Klaten and Soerakarta residencies have been merged into one, making a total of nineteen residencies in the main Java island instead of twenty residencies, as Cribb's map shows. Residencies outside of Java, Sumatra's Weskust, Timor en Onderhoorigheden and Molukken, that showed activities of public meetings are inserted individually within the map.

It is important to note that the GIS maps generated here strictly reflect the data collected from the OV reports in *Sinar Hindia* and *Api* from 1918 to 1926; they do not include other locations in the Dutch East Indies where other communist activities might have been found.[13] The OV reports in *SH/Api* likely did not capture every OV that had been held in or by the *Pergerakan merah*. Some OVs might not be reported in *SH/Api*; some written or oral reports might not have reached the *SH/Api* office and, due to editorial concerns, some reports that arrived at the editorial desk might not be published in *SH/Api*. Based on the statistical analysis, a report could take between a minimum of less than one day to a maximum of 209 days. A length of four days was the mode—the number occurring most frequently—and on average it took 7.5 days for news of a meeting to be reported in the newspaper. Despite this restriction, this data helps elucidate the spatial politics of the early communist movement in the Dutch East Indies, revealing the role of common people and mundane daily political activities as the driving forces behind this era of anticolonial struggles.

The Making of Revolutionary Time and Space

The data reveals for the first time the growth of the movement over time and space, showing that it spread in other islands outside of Java, and that it was mobilized by diverse groups. Map 1 shows the aggregated locations of all OVs from 1920 to 1925, revealing the spread of the movement across Java and in Sumatra Weskust, Timor en Onderhoorigheden (henceforth, Timor), and Molukken residencies. In previous literature, big cities such as Semarang, Surabaya, Batavia, and Surakarta are considered as important locations for the movement.[14] The map expands this geography and shows that the movement was active across many other locations in Java, both in urban and rural areas. It also shows activity in several locations outside of Java, which confirms findings by Audrey Kahin on West Sumatra and Farram on Timor.[15] Within Java, Semarang and Djepara-Rembang residencies that bordered each other were the locations of the most OV held, 180 and 169 meetings, respectively.

Locations of many OVs in Semarang, such as Salatiga, Ambarawa, Weleri, and Kaliwungu, were areas with large plantation companies. Thirty-six plantation companies occupied mountainous towns of Salatiga, Ambarawa, Boja, and Selokaton, while six others were in lowlands Kendal's districts Weleri and Kaliwungu. In Kendal, many of the areas were occupied by sugar plantations. Sugar factories in Cepiring and Gemuk were owned by N. V. tot Exploitatie der Kendalsche Suikerfabrieken, while a plant in Kaliwungu was owned by Kaliwungu-Plantaran plantation company, consigned with Cultuurmaatschappij der Vorstenlanden. Operating the sugar factories led to an increased demand for lands to ensure expansion of plantation areas and access to irrigation. These companies would need to lease lands from indigenous Indonesians, and conflicts were often unavoidable. OVs were often held in these areas to mobilize public support on communal demands, such as the return of rice paddy fields and embankments to villagers.[16] As a trade and port hub in Central Java, Semarang was well connected with train and tram traffic, which was managed by Nederlands-Indische Spoorweg (NIS) and its branches Semarang-Cheribon Stoomtram Maatschappij (SCS) and Semarang-Djoewana Stoomtram Maatschappij (SJS).[17] Cars and buses were also available, and the number of transportation options created a well-connected transit network alongside a myriad of job opportunities. As a result, new workers kept pouring into the cities, including immigrants from Europe and China.[18]

The neighboring residency with the second highest number of reported meetings, Djepara-Rembang, was much different than Semarang. Rembang's land was barren, dry, and difficult to cultivate. Its soil was composed of barren limestone, which made it an ideal site to grow teak but a poor site to grow other crops. A large area of Rembang residency was planted with Government teak forests.[19] No European plantation companies operated in Rembang, let alone profitable sugar companies. Rembang relied on crops—peanuts, corn, tobacco, coconuts, kapok, and chilis—to be both consumed and exported, and the residency also produced batik (traditional textiles). Fisheries along the coast of Rembang also provided income to the locals, but Djepara-Rembang was relatively impoverished compared to Semarang.[20] Transport access reflected this as well. Trams and trains were provided by the Semarang-based NIS and SJS. Roads along railways connecting Bojonegoro-Cepu-Blora and Rembang-Lasem were in good condition, but beyond them the traffic network was relatively inaccessible for common people. The available roads were primarily used to transport heavy materials and forest products, and the southern areas of Madiun and Kediri remained isolated because of a lack of roads.[21] Despite this, communism took root in this area, including among Cina Peranakan (Chinese descendants). Some of these Chinese communist leaders were arrested after the 1926 to 1927 revolts and six of them were sent to Boven Digul in Papua.[22] Semarang and Djepara-Rembang were vastly different in terms of economic development, but both still served as centers of most communist OVs.

Maps 2 through 4 show that, within a brief period between 1920 and 1925, the number of reported OVs held in the Indies quadrupled and expanded into new areas creating new centers of the movement. The number of OVs increased fourfold from seventy-three, forty-one, thirty-five, and thirty-seven meetings held between 1920 and 1923 to 296 meetings in 1924 and 344 meetings in 1925. While table 4 indicates the highest numbers of OV in the years 1924 and 1925, it does not provide any other information regarding the spread of the movement. In contrast, maps 2, 3, and 4 allow us to see how the movement expanded from Central Java to West Java in 1923 and then to East Java and a couple of the Outer Islands by 1925. While Semarang and Djepara-Rembang continued to be important centers of OVs, this aggregated number shows us that, in 1924 to 1925, the movement spread to new centers across Java, including Batavia, Preanger, Madioen, Kediri, Malang, Besoeki, and in the Outer Islands, such as Sumatra Weskust, Timor, and Molukken.[23] Centers in the eastern part of Java, especially

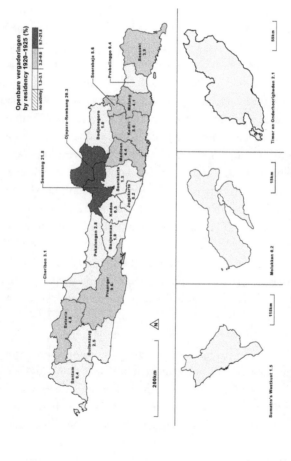

Openbare vergaderingen
by residency 1920–1925 (%)

no activity | 1.3–3.1 | 3.3–9.6 | 9.7–21.8

Semarang 21.8
Djapara–Rembang 20.3
Bodjonegoro 1.8
Malang 4.1
Kediri 3.6
Madioen 4.4
Soerabaja 5.6
Probolinggo 8.4
Basoeki 3.3
Soerakarta 1.3
Djogjakarta 0.2
Kedoe 0.5
Banjoemas 1.0
Pekalongan 2.8
Cheribon 3.1
Preanger 9.6
Batavia 4.8
Buitenzorg 2.5
Bantam 0.4

200km

Sumatra's Westkust 1.5
115km
Molukken 0.2
15km
Timor en Onderhoorigheden 2.1
55km

MAP 1.   *Openbare vergaderingen* by residency in 1920 to 1925.

76

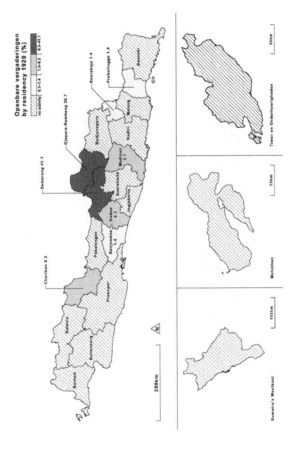

MAP 2. *Openbare vergaderingen by residency in 1920.*

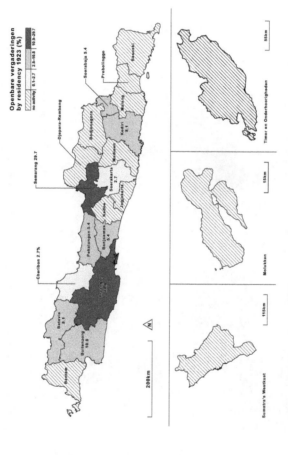

MAP 3. *Openbare vergaderingen* by residency in 1923.

Openbare vergaderingen
by residency 1923 (%)

no activity | 0.1–2.7 | 2.8–10.8 | 10.9–29.7

Cheribon 2.7%

Samarang 29.7

Djapara-Rembang

Bodjonegoro

Preanger
Samoem
Batavia 3.1
Buitenzorg 10.8
Pakalongan 5.4
Kedoe
Banjoemas 5.4
Madioen
Kediri R.t
Malang
Soerakarta 2.7
Jogjakarta
Soerabaja 5.4
Probolinggo
Besoeki

200km

Sumatra's Westkust   115km

Molukken   15km

Timor en Onderhoorigheden   55km

78

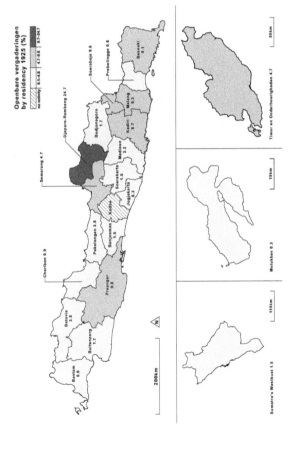

Openbare vergaderingen by residency 1925 (%)

no activity | 0.1–4.6 | 4.7–9.6 | 9.7–24.7

Samarang 4.7

Djapara-Rembang 24.7

Bodjonegoro 1.7

Soerabaja 9.6

Probolinggo 0.6

Besoeki 6.1

Madioen 3.2

Kediri 6.7

Malang 9.3

Soerakarta 1.5

Jogjakarta 0.3

Kedoe

Banjoemas 1.5

Pakalongan 3.8

Cheribon 0.9

Batavia 3.8

Buitenzorg 1.7

Bantam 0.6

Preanger 9.0

200km

Sumatra's Westkust 1.5   115km

Moluksen 0.3   15km

Timor en Onderhoorigheden 4.7   55km

MAP 4. *Openbare vergaderingen by residency in 1925.*

79

TABLE 4. Number of *vergaderingen* (meetings) reported in 1920 to 1925

LOCATION	1920	1921	1922	1923	1924	1925
Bantam	0	0	0	0	1	2
Batavia	0	1	0	3	23	13
Buitenzorg	0	2	0	4	9	6
Preanger	0	2	1	8	37	31
Cheribon	6	6	2	1	8	3
Pekalongan	0	2	3	2	3	13
Semarang	30	20	15	11	88	16
Djepara-Rembang	29	5	6	0	43	85
Banjoemas	1	0	0	2	0	5
Kedoe	3	0	0	0	1	0
Jogjakarta	0	0	0	0	1	1
Soerakarta	0	0	2	1	3	5
Madioen	2	1	4	0	18	11
Kediri	0	0	1	3	37	30
Malang	0	1	1	0	0	32
Soerabaja	1	1	0	2	9	33
Probolinggo	1	0	0	0	0	2
Besoeki	0	0	0	0	4	28
Bodjonegoro	0	0	0	0	2	6
Sumatra's Westkust	0	0	0	0	7	5
Timor en Onderhoorigheden	0	0	0	0	1	16
Molukken	0	0	0	0	1	1

Kediri, Soerabaja, and Malang, were sites of sugar factories and plantations.[24] The relationship between workers in the sugar industry gives clues to the social composition of the movement; plantation workers and peasants were equally important movers of the movement, as were the transport and communication workers that were discussed in chapter 1. These maps show that movement developed within five years, quickly spreading through new areas.

The 1923 Ban of *Openbare Vergaderingen*

Two events in 1923 led the *pergerakan merah* to become more popular and more radicalized: the split within SI and the ban of OVs. These events happened at the same time as the arrest and exile of an important party leader named Semaoen. His banishment, along with that of other party founders who previously had been sent to exile abroad, including Sneevliet, A. Baars, and Tan Malaka, left the movement in the hands of rank-and-file members and relatively new party leaders, such as Soemantri, Darsono, and Subakat, who would also soon face jail time and expulsion. While 1923 was a turning point in the movement, 1924 and 1925 were the most radical years of the movement and were mobilized by local members.

The movement was largely inactive from 1920 through early 1923. Following the postwar economic recession of 1918 to 1920, trade unions collapsed, and the PKI nearly lost most of its followers; its total number of participants shrank to 200, and most of these members lived in big cities. In response, communist courses in more populous areas encouraged trade unions and other political movements to unite in the so-called Radical Concentration. The PKI and Red Sarekat Islam (later transformed into Sarekat Rajat, SR) were the leading organizations in Radical Concentration, which helped SR quickly spread throughout Java. As SR established sections in local areas, each section was enrolled as a branch of the PKI while continuing to operate under the banner of SR. A communist member would then provide express courses and teach PKI's program to the members of the branch.[25]

While this partnership was unfolding, in October 1921, the alliance between the PKI and SI began to break. The central leadership of SI decided to install "party discipline" by purifying its uniting principle under Islam instead of communism.[26] They saw that the communists had infiltrated and made use of SI's followers, and the connection with the PKI gave SI a bad reputation. This was soon to be followed with a break within SI into what became "SI merah" (Red SI) and "SI putih" (White SI). Red SI joined the PKI in mobilizing as the *pergerakan merah*. The name would later be turned into Sarekat Rajat (People's Union, SR) removing the term Islam as it purportedly countered the goal of uniting all elements of the Indies against colonialism. After the general strike led by the VSTP in 1923, which will be detailed later in this chapter, membership in SI declined considerably, and, by the end of the year, most of the small traders and peasants who dominated SI was absorbed into its revolutionary wing. By the beginning of 1924, the communist party led the popular following of both SR and labor unions and became the only popular organization mobilizing a revolutionary cause against Dutch imperialism.[27]

As most popular followers joined the communist cause, one of the biggest strikes was held in Semarang, led by the VSTP. As discussed in chapter 1, founded in 1908,[28] the VSTP was one of the oldest and most progressive Indonesian labor unions, and was known as a "powerful and best-led organization."[29] Some of the famous leaders in the history of Indonesian communism began their careers there before establishing the PKI. The VSTP's members consisted of skilled and semi-skilled Dutch and Indonesian urban railway workers, and together they created a sector of an educated working class. Though centralized in Semarang, the VSTP's branches were spread all over Java, especially along the railway lines. This explains how the *pergerakan merah*, of which the VSTP was a part, spread out to other cities and the hinterlands across Java. The 1923 strike illustrates this process.

On January 1, 1923, the Dutch colonial government's decision to reduce the cost-of-living bonus took effect. This announcement was followed by the private rail line's statement that they would also implement major wage and personnel cuts. At the VSTP's congress in February, the workers agreed to strike.[30] The plan to hold the strike continued to be discussed in OVs held by either the VSTP or SI over the next several months. In early May that year, as the strike plan progressed, Semaoen was accused of planning a threat to public order and was arrested. His arrest provided the momentum for VSTP members to hold the strike. On May 9 through 11, 1923, strikes occurred in train stations in the western part of Java in Tjikampek and Cheribon, in Central Java in Ambarawa, Tegal, Semarang, Jogja, Solo, Pekalongan, and Kudus, and in the eastern part of Java in Soerabaja, Bangil, Malang, Klakah, Pasoeroean, Madiun, Gubeng, Kertosono, Wonokromo, Sidhoardjo, Ponorogo, Kroja, and Djombang, halting transportation across Java to a dead stop.[31]

The success of the strike lay in its well-structured and organized planning. One leader emphasized that the strike "must be well disciplined and properly timed, and must not consist of local ventures and wildcat walkouts and sabotage."[32] The strikes that began with railways workers in trains and stations were soon followed by strikes in workshops and factories as other professions joined the fight.[33] The May 1923 strike would become the biggest strike ever held in its time. It caused many dismissals among the strikers, who in subsequent years would become leaders of SR. In response, the Dutch government sent Semaoen into exile; he was forced to leave his wife, Salemah, and their newborn child (Semaoen's venture abroad is discussed further in chapter 7).[34] The government also arrested many leaders and members and revoked the rights to hold *vergaderingen*. The ban was installed not at the central level by the governor-general but at the level of residencies, starting in the residency of Semarang and subsequently to the neighboring residencies. It was applied not just to communist-related organizations but for all native and nonnative organizations, creating an uproar of discontents even among noncommunist activists and newspapers. The revocation of meeting rights was not lifted until October that year, which was long enough to sway some public opinion toward the communist cause.

These two events drew a popular following from the lower-class population, and the movement's expansion can be observed in maps 5 and 6. After the 1923 strike, the number of *vergaderingen* swelled fourfold, from 153 before the ban to 673 after it. Some of the residencies that witnessed a dramatic increase in

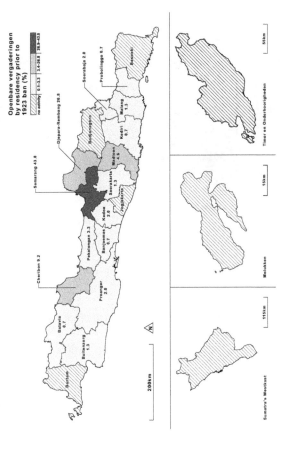

Openbare vergaderingen by residency prior to the 1923 ban (%)

no activity | 0.1–3.3 | 3.4–26.8 | 26.9–43.8

Samarang 43.8
Cheribon 9.2
Batavia 6.7
Bantam
Buitenzorg 1.3
Preanger 2.0
Pekalongan 3.3
Banjoemas 6.7
Kedoe 2.8
Djokjakarta
Soerakarta 1.3
Madioen 4.6
Kediri 6.7
Bodjonegoro
Djapara-Rembang 26.8
Soerabaja 2.0
Probolinggo 0.7
Malang 1.3
Besoeki

200km

Sumatr'a Westkust 115km
Molukken 15km
Timor en Onderhoorigheden 55km

MAP 5. Openbare vergaderingen by residency prior to the 1923 ban.

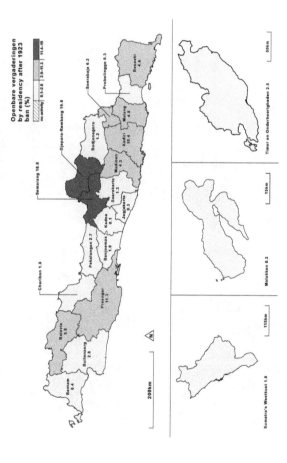

MAP 6. *Openbare vergaderingen* by residency after the 1923 ban.

Openbare vergaderingen
by residency after 1923
ban (%)

no activity 0.1–2.8 2.9–11.3 11.4–19

Samarang 16.8
Djepara-Rembang 19.0
Soerabaja 6.2
Probolinggo 0.3
Besoeki 4.4
Bodjonegoro 1.2
Malang 4.8
Kediri 10.4
Madioen 4.3
Soerakarta 1.3
Jogjakarta 0.3
Kedoe 0.1
Banjoemas 1.0
Pekalongan 2.7
Cheribon 1.6
Preanger 11.3
Batavia 5.8
Buitenzorg 2.6
Bantam 0.4

N

200km

Sumatra's Westkust 1.8
115km

Molukken 0.3
15km

Timor en Onderhoorigheden 2.5
55km

the number of meetings were Batavia (from one to thirty-nine), Preanger (from three to seventy-six), Semarang (from sixty-seven to 113), Djepara-Rembang (from forty-one to 128), Madioen (from seven to twenty-nine), Kediri (from one to seventy), Malang (from two to thirty-two), Soerabaja (from three to forty-two), and Besoeki (from zero to thirty-two). The Outer Islands, including Sumatra's Westkust, Timor, and Molukken, all held communist *vergaderingen* for the first time after the 1923 ban. After the revocation of the rights of assembly, the movement became more attractive and more active in expanding its influence across geographical locations.

Since 1923, according to a report to Comintern, the PKI had become the most important party in the Indies. By 1920 nationalist parties were practically disorganized and most of their revolutionary members turned to the PKI.[35] By the end of 1924, the communist party "was growing in every direction in stronger manner as it had ever been before, partly aided through the continuous decay of Sarekat Islam." The PKI's key tactic was revolutionizing intellectuals, creating Left Wings in national organizations such as Boedi Oetomo. This occurred as the government raised taxes by 3,000 percent, impoverishing much of the middle and lower classes.[36] In 1925, restrictions around communicative practices and actions such as publication and assembly continued to be implemented (see chapter 6). Provocations organized by the police continued to occur and arrests of editors and organizers happened much more frequently. "These provocations compromise the police to such an extent that even many Dutch newspapers condemn them." While the PKI and its affiliated organizations worked alone and independently, it continued to receive sympathy from other national organizations. Together with other national organizations, they formed a bloc to place "democratic demands before the government."[37] With support from national organizations and rapid increases in membership, the years from 1924 to 1925 were halcyon days of the early Indonesian communist movement. The movement would face a temporary demise just one year later.

Spatial and Social Characteristics

The precise social composition of the movement is difficult, if not impossible, to grasp. While we have access to membership numbers in some official reports, many people joined the movement without obtaining formal training or membership. This becomes apparent when we delve into the qualitative day-to-day experiences of living in the movement. It is wise to take numbers in formal documents as the lowest possible calculation. The Indonesian delegates' statement that "our influence is 10 times greater than number of the enrolled members" is not unreasonable.[38] Report of these PKI representatives to the Comintern shows that, after 1923, enrolled members of SR amounted to 100,850 people, comprising of 35 percent workers, 55 percent peasants, and 10 percent petty capitalists and semi-intellectuals.[39] Trade union membership totaled 23,195, 70 percent of whom joined the revolutionary movement.[40] The PKI boasted 8,000 members, comprised of 30 percent factory workers, 42 percent semi-intellectuals—small bureau officials and government employees, 0.1 percent big intellectuals, and 20 percent petty traders and craftworkers, and the remaining 7.9 percent peasants.[41] It is clear that while most PKI members were factory workers and white-collar workers, the movement's membership from the rank of the peasants came mostly from SR.

The PKI and SR also held a ten-month course on communism for its members. The list of these members shows the detailed numbers of members, and their different cities and residencies. Residencies with the most members who took the course were relatively the same as those with the most OVs: Bantam (125 PKI members and 10,000 SR members), Bandoeng or Preanger (965 and 14,350), Semarang (750 and 13,250), and Malang (200 and 11,850). However, table 5 gives valuable information about courses taken by members outside of Java that the OV maps do not capture. These courses were held in Atjeh, Medan, Padang, Palembang, Menggala, Giham, Pontianak, Bandjarmasin, Makassar, Gorontalo, Ternate, and Timor.[42]

The movement mobilized the army and the police. In the army, they aimed to influence the sergeant; each sergeant oversaw sixty men, so if they could influence five sergeants, they would receive up to 300 members. Before the OV ban, twelve to twenty-five meetings were held for the army; however, only the first-class soldiers were able to attend due to the midnight curfew. First class soldiers refer to those who did not "make any mistakes, who did his duty well and after a period of service of 12 years, they get the first rank. The first rank soldier gets greater freedom and more wages."[43] The PKI intensively worked on mobilizing the army and the police because the majority of them were native Indonesians who would sympathize with the anticolonial cause and because of their strategic roles and access to arms and weapons.[44]

While these numbers help us imagine the growth and composition of the movement, they could not provide valuable information on the identity of the followers. OV GIS maps provide us insights for the first time regarding the daily process of mobilization and the involvement of women and people of different races.

The OV reports often included the number of people who attended the meetings. This information would specify the gender and the nationalities of the attendees. For example, in an OV held by SI Semarang on May 8, 1921, the report said, "4,000 people attended, also Baars [a Dutch communist leader who

TABLE 5. Number of PKI and SR members who took the 10-month communist course by location (city)

RESIDENCY	PKI	SAREKAT RAJAT
Batavia	635	6,550
Bantam	125	10,000
Bandoeng	965	14,350
Tjirebon	250	1,400
Tjilatjap	120	700
Magelang	130	2,200
Pekalongan	425	6,000
Semarang	750	13,250
Djokdjakarta	475	2,000
Solo	195	4,000
Tjepoe	300	3,000
Madioen	230	2,400
Kediri	680	5,500
Soerabaya	525	6,000
Malang	200	11,850
Banjoewangi	155	2,500
Atjeh	80	400
Medan	105	500
Padang	370	2,300
Palembang	50	150
Menggala	70	500
Giham	15	100
Pontianak	30	100
Bandjarmasin	10	500
Makassar	75	500
Gorontalo	25	100
Ternate	980	3,000
Timor	30	1,000
Total	8,000	100,850

co-founded the ISDV], and also Dutch and Chinese people as well as women."[45] Based on the number of the attendees that I classified into several groups, that are "500 and fewer," "500–1,000," and "1,000 and more," most meetings were attended by 500 people or fewer.

Over the years, more meetings were held and more participants were active, but the individual meetings had fewer attendees. Approximately 39.4 percent of *vergaderingen*, or a total of 341 meetings, were attended by 101 to 500 people, and 32.3 percent, or a total of 279 meetings, were attended by fewer than one hundred

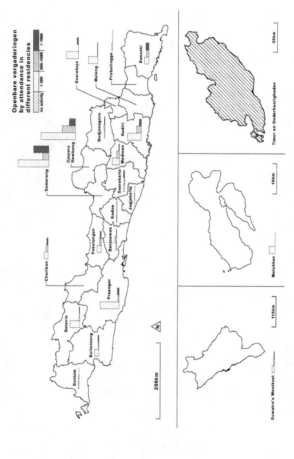

MAP 7. *Openbare vergaderingen* by attendance in different residencies 1920–25.

people. Additionally, 14.1 percent of meetings were attended by 1,000 to 10,000 people. Out of that 14.1 percent, ninety-eight meetings were attended by 1,001 to 3,000 people, nineteen meetings by 3,001 to 5,000 people, and five meetings by 5,001 to 10,000 people. Around 71.7 percent of the total *vergaderingen* were attended by 500 people or fewer. This is likely because the meetings were usually held in houses; if they were held in remote areas, big rooms that could hold more than 500 people might not be available, and OVs could not be held in an open space area like parks (see chapter 6). As shown in map 7, the meetings with the number of attendees that exceed 1,000 people took place in Semarang and Djepara-Rembang residencies, both large port cities. By contrast, meetings in other residencies were usually attended by fewer than 500 people.

A correlation run between the number of attendees and the sponsoring organization also demonstrates that meetings held by red wing (communist-leaning) organizations—Red Islamic Union, People's Union, and PKI—tended to be attended by fewer participants (449 people) than those sponsored by SI (836 people) before the name changed. The communist meetings were attended by a much smaller number of people, especially after the government put communism under its close watch, further revealing that the meetings were likely no longer held by main leaders who were by then under government surveillance. Instead, it suggests that the meetings were held in houses belonging to ordinary people, oftentimes in remote, rural areas where large office buildings were not available. On the other hand, before the government ban and the split of the party, SI meetings were attended by a larger number of people and could occur more openly in office buildings and halls.

As the movement became more radicalized and more popular and the meetings became more frequent yet smaller, the composition of the people attending also became more diverse. Map 8 demonstrates that women attended most *vergaderingen* except those in Kedoe, Probolinggo, and Timor residencies. Semarang, Djepara-Rembang, Kediri, Soerabaja, Besoeki, Malang, Preanger, and Batavia residencies were the locations of at least twenty meetings in which women participated. Based on a correlation test run between women's attendance and year of event, we can conclude that women began to attend OVs more often from 1924 onward. This strengthens the finding that the movement expanded in both size and diversity after the 1923 ban.

In the meantime, while people of other nationalities who lived and worked in the Indies—Dutch, European, Indian, Arab, and Chinese—often attended the meetings, Chinese members still dominated this group. Eighty-five meetings were attended by Chinese members, compared to fifteen by Europeans and twelve by people of other nationalities (Indian and Arab). The *pergerakan merah* worked extensively with the Chinese (about two million Chinese immigrants lived in

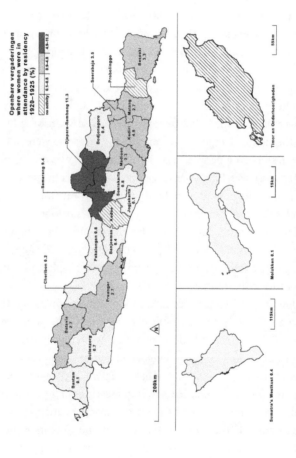

MAP 8. *Openbare vergaderingen* where women were in attendance by residency 1920–25.

Java alone). During the Shanghai strike in 1925, OVs would collect funds for the strikers, creating mutual desire among the Chinese to work with the movement. Cases of discrimination against them by the Dutch colonial state would also often be discussed in OVs. As a result, during the repressions against the movement, "practically all the Chinese papers sympathized with [the communists] and protested against this oppression."[46] The cooperation with the Chinese was not just a matter of raising solidarity; it was also a strategic call to organize a center in China. In the face of daily oppressions and the exiles of many leaders, having a center to represent the need of the Indonesian people located between Russia and the Indies would help "strengthen the Party Central Committee inside the country."[47]

The involvement of women and nonindigenous people in OVs discloses two important points. First, unlike histories that often see anticolonial resistance as the East or the colonized against the West or the colonizers, the Indonesian anticolonial movement was clearly supported by people of other nations, including people from colonized nations. The involvement of Dutch people alongside native Indonesians, Chinese, and other Asians shows that the battle did not exclusively belong to the indigenous people. However, the reverse is also true: Indigenous people were often supporters of colonialism. Resistance in non-Western societies against colonialism was not something triggered only by their identity as "non-Western"—it was not merely the responses of "indigenous Eastern" people against "Western" imperialism. With the involvement of people of other nationalities in the early communist movement in this period, the movement spoke of anticolonial resistance more as shared common interests across nationalities rather than simply as one indigenous political unity. Additionally, the involvement of women and of nonnative nationals indicates that the movement built solidarity across different sexes and nationalities.

I gathered the names of speakers and chairpersons from the *vergaderingen* reports (see appendix) and processed them based on the frequency they spoke in and/or led a meeting. Based on over 900 names of speakers and chairpersons, I found that about 69 percent spoke or led in one to two meetings, 18 percent in three to five meetings, 7 percent in six to ten meetings, 5 percent in eleven to twenty meetings, and fewer than 1 percent in twenty-one to thirty-eight meetings. It is likely that the propagandists were not just the people sent by the communist-related parties from the headquarter in Semarang, but instead were composed of local ordinary people leading the debates and discussions.

I then processed the data differently by regrouping the names of speakers into "popular speakers." Popular speakers referred to those who spoke in at least ten meetings; nonpopular speakers were those who spoke in less than ten meetings. I came up with a total of fifty-nine names of popular speakers who spoke

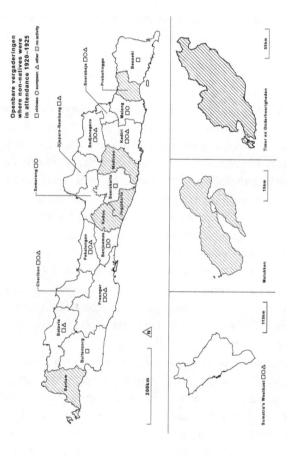

MAP 9. *Openbare vergaderingen where nonnatives were in attendance 1920–25.*

in 18.4 percent of the meetings. The number of meetings that were led by non-"popular speakers" was four times more than those led by the popular ones. This confirms the previous finding: The speakers of the meetings were not dominated by popular leaders, which further demonstrates that the meetings were participatory and democratic.

While the speakers were diverse, some speakers moved around and women took on more leading roles. From a total of seventeen meetings a communist leader named *woro* Moenasiah led or spoke at, these meetings were held in three residencies, Semarang, Djepara-Rembang, and Kediri. *Woro* Moenasiah lived in Semarang and, with easy access by train to Djepara-Rembang and Kediri, she could go to these neighboring residencies. By comparison, Semaoen traveled across the island of Java, reaching as far as Batavia and Cheribon in West Java, Pekalongan, Djepara-Rembang, and Semarang in Central Java, and Soerabaja in East Java. Compared to the movement of *woro* Moenasiah, Semaoen traveled to more distant areas than she did.

I found evidence of ninety-one women who spoke and led public meetings, and thirty-eight spoke in a meeting more than once. Women speakers spoke at 198 meetings held in most residencies in Java and the Outer Islands, except in Cheribon, Banjoemas, Jogjakarta, Probolinggo, and Timor. In Java, Semarang, Djepara-Rembang, Soerabaja, Kediri, and Batavia—all big cities—were the residencies where at least ten meetings were held that had women speakers. Outside of Java, Sumatra's Weskust held two meetings, and Molukken held one meeting with women speakers.

After the 1923 ban, people of other nationalities, primarily those of Chinese origin, joined as speakers. I located the names of seventeen Chinese people who led and spoke in *vergaderingen* held in Buitenzorg, Preanger, Pekalongan, Semarang, Banjoemas, Kediri, Malang, Soerabaja, Besoeki, and Bodjonegoro. Leo Suryadinata notes that Kho Tjun Wan and Tan Ping Tjiat were among the two prominent peranakan Chinese in the PKI.[48] Kho Tjun Wan led a meeting in Semarang on January 29, 1925, along with a number of speakers, including *woro* Moenasiah and *woro* Soetitah, attended by 1,000 people—among them Chinese—to express grievances against Semarang city administration. Likewise, Tan Ping Tjiat spoke in two meetings. The first one was at Liem Tjeng Kwee's house in Kampoeng Kradjan Petjinan on August 2, 1925, which was attended by 1,200 "women and men from various races and religions" to discuss, along with a communist leader Moeso, the revolt in China and Morocco, the role of students in the revolt in China, and the comradeship between native Indonesians and Chinese people. The second meeting was in Bodjonegoro on September 19, 1925, at Hadji Mashadi's house, and was attended by 1,200 Indonesians and Chinese.[49] Outside of Java, no speakers of other nationalities were involved, but within Java, as the *pergerakan merah* grew, these people of different sexes and nationalities took greater leadership roles in the movement.

Collective Organizations

A study on the names of the organizations holding the meetings indicates that this period was mobilized by a network of several different organizations. These include "Sarekat Islam" (SI), "Sarekat Islam Merah" (Red Islamic Union, SI Merah), "Sarekat Rajat" (SR), "Partai Komunis Indonesia" (PKI), "Sarekat Islam Perempuan" (Women's Islamic Union), "Sarekat Rajat Perempuan" (Women's People's Union), and various labor unions. Before the split of the Red and White wings of SI, all branches of SI bore the names "SI" only. After the split that was completed in 1923, the communist-leaning branches changed their names into "SI Merah" and then "SR."

Apart from Preanger and Semarang, most meetings were held by communist-leaning organizations: SI Merah, SR, and the PKI. In fact, out of the total meetings reported, SI meetings comprised only 19 percent. The rest were run by the communist organizations. Within the communist organizations, the number of OVs held by the communist SR was the highest, at 44 percent, and meetings held by the PKI, the communist party, was actually low, at only 8.7 percent. In other words, the PKI was not the main organizer of OVs.

Because this movement was mobilized by a united front of different organizations, the people of the period called the movement *"pergerakan merah."*[50] To confirm the limited role of the PKI in mobilizing *vergaderingen*, I ran GIS mapping on the number of *vergaderingen* held only by the communist-leaning organizations, including SI Merah, SR, and PKI. Map 10 reveals that most meetings were held by the former SI branches, which by then bore the new names SI Merah and SR. Except for Surabaja and Besoeki, which became the basis of PKI meetings, my findings suggest that the *pergerakan merah* in the Indies was not mobilized by the PKI alone; instead, its success is owed to communist-leaning SI networks. In other words, while PKI membership was small and centered around a small group of leaders and workers, the massive following of the movement came from the SI branches that had shifted its direction by uniting with communism. This confirms further that the movement was not a product of centralized PKI propaganda. Instead of calling it the period of the PKI, I opt for the *"pergerakan merah"* (the red movement) to describe the network of independent yet interrelated parties and unions that mobilized the anticolonial struggles under the banner of communism.

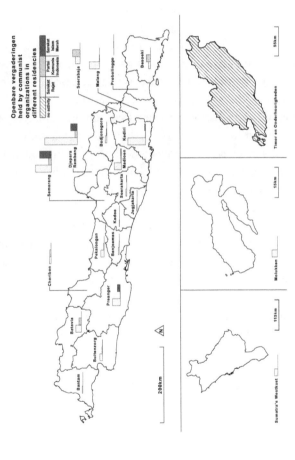

MAP 10. *Openbare vergaderingen* held by communist organizations in different residencies

Mental Landscapes of the *Pergerakan Merah*

Another way to explain whether the movement was mobilized democratically from the bottom up or hierarchically from the top of central communist party to the bottom is by looking at the topics discussed in *vergaderingen*. If topics sent from the formal central party dominated the meetings, this could indicate that the meetings were held to expand the formal parties' interests. I classified the topics into nine groups: analysis, organization, news, women, *reactie* (reaction), school, local matters, Islam, and law. "Organization" consists of topics related to formal parties, including vision, mission, structure of the organization, votes on leaders, and party motions. Biographies of leaders were often discussed here too, including locals like Tan Malaka, Semaoen, H. Datoek Batoeah, and N. Zainoeddin from Sumatra, and international leaders such as Sun Yat-Sen, Karl Radek, Leon Trotsky, Joseph Stalin, Grigory Zinoviev, Leonid Krassin, and Mikhail Kalinin. "Analysis" includes broad discussions on capitalism, communism, class struggle, freedom to mobilize and to write, and oppression and exploitation. In "News," people discussed wars that were occurring at the time, news from Russia, China, and other colonized countries, reports on labor strikes in other countries, and local news on the movement. "Women" involved a discussion on women's emancipation, women's place in the movement, and the specific types of oppression and exploitation. "*Reactie*," one of the important vocabularies of the movement, involved a discussion on the countermovement and actions led by people who did not agree with the movement. From this discussion, we can understand how anticommunist propaganda in Indonesia emerged for the first time. "School" includes discussions on education in capitalist society, the creation of SI schools, education in general, and the place of Islamic schools in the movement. "Local matters" is a specific discussion on problems related directly to the attendees' living conditions, from lack of rice production, broken bridges, lack of train access, village security, factories versus rice fields, to the cost of hajj (going to a pilgrimage to Mecca). "Law" comprises of the discussions on the changing policy and regulation affecting the movement. In "Islam," people debated the compatibility of the religion with communism and how important it was to be a revolutionary Muslim.

Table 6 provides information on how the mental landscape of the *pergerakan merah* comprised of varied topics that were discussed and debated, and these topics carried forth aspirations of red enlightenment. Based on the aggregated number of the topics, "Analysis" and "Organization" were most frequently discussed in the meetings; about 400 meetings discussed these topics. "News" came next, with 196 meetings that included discussion of international and domestic reports. Ninety-two meetings talked about women's issues in the movement;

TABLE 6. Topics discussed in *openbare vergaderingen*

FREQUENCY	DEFINITION
400	Analysis
395	Organization
196	News
92	Women
89	*Reactie*
86	School
78	Local matters
50	Islam
21	Law

eighty-nine discussed anticommunist propaganda; eighty-six talked about the importance of education for communist cadres—children and adults; seventy-eight talked about local and immediate local problems the people faced; fifty focused on Islam; and twenty-one on law. Any given meeting could include a discussion of several of these topics. The graph shows that while "Analysis" and "Organization" remained the important topics debated in the meetings, other topics on news, women, *reactie*, and education were also deemed important. Through meetings, the *pergerakan merah* acted like an educational institution in which matters regarding organization and collective issues were debated and discussed daily, which shows how the movement embodied the project of red enlightenment. The breadth of the topics discussed in the meetings reveals that the widespread scope of the movement was manifested both geographically and in the topics people discussed.

Common Spaces: The Blurring of Public/Private Boundaries

An examination of the spaces where the meetings were held reveals the blurring boundaries of private versus public spaces. As the movement became more popular, private spaces were politicized to hold public discussions of shared common concerns and topics discussed in this chapter. Likewise, public spaces were occupied to discuss private issues that often were turned into the agenda of the movement, like women's emancipation and the reformation of Islam. The OVs were held in various spaces, including "private house" (283) "village" (135), "office" (78), "cinema" (56), "school" (47), "private building" (26), and "*stamboel/* art theater" (7). When a report says *desa* (village), the OV was most likely held in a house instead of *balai desa* (village meeting hall), given that not all local government was in support of the *pergerakan merah*. This means that the meetings in "villages" most likely occupied private houses. The combined number of "village" and "private house" would then generate a figure of 66 percent in which meetings were held in private homes.

While private houses could only host a few hundred attendees, meetings with a larger number of participants were held in places like cinemas, schools, *stamboel* or art theaters, and offices. In cinema spaces, the number of the participants could range between 3,000 to 10,000 people. Map 11 is useful, as it shows us the existence of cinemas, many of them owned by peranakan Chinese, all over Java and how movie theaters became the main locations that could offer space for large numbers of attendees.

Besides cinemas, people also used school buildings for meetings. These were SI schools (later People's Schools) built in recent years by the *pergerakan merah* (see chapter 4 for more discussion on SI schools). The use of SI school buildings as spaces for public meetings reveals the widespread existence of SI schools, as map 11 shows.

This map shows that SI schools existed in Bantam, Buitenzorg, Batavia, Preanger, Cheribon, Pekalongan, Semarang, Madioen, Kediri, Bodjonegoro, Soerabaja, Besoeki, and Sumatra's Westkust. This is not an exhaustive list, and further data from other sources are needed to have a more complete picture of the existence of SI schools.

The next two spaces that could accommodate greater numbers of attendees were *stamboel*. Map 11 shows the existence of *stamboel* in various parts of Java: Cheribon, Preanger, Pekalongan, Semarang, Djepara-Rembang, and Malang. *Stamboel* was an important and popular kind of theater at the time, and this map reveals that the meetings were highly scattered.[51]

The political parties usually held meetings in their offices, and the above map shows the existence of those offices across Java, in Sumatra's Westkust, and in Molukken. This map shows cinemas, offices, schools, and *stamboels* as "spaces of contention" and indicate the movement's mundane and ordinary quality. OVs could not be held in such a scale without occupying and repurposing existing spaces, as they required a congregation in a designated place. While a propagandist needs a follower and a shared political cause, this could not happen without a space and place of contention. In the Indonesian communist history, unlike formal party congresses that took place in halls and office buildings, much of the *pergerakan merah* occupied ordinary spaces in people's lives, turning private spaces for dwelling into public ones for education, entertainment, and organizing. It was public not just in terms of being open to everyone, but also in that it created shared communal or common spaces.

Openbare vergaderingen location type by residency 1920–1925

□ cinema ○ SI school building △ stamboel
▨ no activity □○△

Cheribon □○△

Semarang □○△

Djapara-Rembang □△

Batavia □○

Buitenzorg □○

Bantam ○

Pekalongan □○△

Preanger □○△

Banjoemas □

Kedoe □

Soerakarta △

Djogjakarta ○

Madioen ○

Bodjonegoro □

Kediri □○

Malang △

Soerabaja □○

Probolinggo

Besoeki □○

200km

N

Sumatra's Westkust ○   115km

Molukken   155km

Timor en Onderhoorigheden   55km

MAP 11. *Openbare vergaderingen* location type by residency, 1920–25

Police State: A Spatial Perspective

Another important aspect of OV reports is that they contain information on whether police watched the meetings. The following map shows the breadth and coordination of police surveillance. According to the reports, police watched most meetings, with the notable exclusions of Probolinggo, Timor, and Molukken. Most of the police observations occurred in meetings held in Preanger (forty-one meetings), Semarang (fifty-one), Djepara-Rembang (sixty-four), Madioen (thirteen), Kediri (thirty-five), Malang (twenty), Soerabaja (twenty-six), and Besoeki (twenty-two). Police watch became more intense in the years after the ban of OVs in 1923 but especially in the years of 1924 and 1925 when the number of OVs held was the highest. This opposition likely triggered the increase in the number of OVs held in those years, but it was also the reason why, by the beginning of 1926, the number of OVs suddenly died out. Anticommunist violence intensified and discouraged ordinary people of the *pergerakan merah* to attend and hold meetings.

Investigating geographic factors deepens our understanding of individual communists' experiences and the nature of legal actions that led to the suppression of the *pergerakan merah* and the subsequent movements of that nature. The repression of public meetings and the extent of anticommunist harassment prior to the communist revolt in 1926 will be discussed further in chapter 6.

A Window into the Past

The GIS maps of *vergaderingen* are a window for us to look at the mundane processes of the making of the *pergerakan merah*. They also provide information on the mental, social, spatial, and political composition of the movement via a visual representation of the exact mapping of the movement and the social composition of its members. The visual representation is an important and innovative finding in its own right, as it reveals a vital point about the making of solidarity. As we have seen, solidarity was built not just across widespread geographical locations, but also across different identities, including women and people of other nationalities. The movement was mobilized not through a leader-centric or a party-centric development, but through a democratic participation of ordinary lower-class people. Democratic participation was formed not through an abstract, static, isolated, and homogenous form of class struggle, but through the inclusion of people from different regions, nationalities, religions, and genders. By 1924, the *pergerakan merah* was the sole mobilizer of common people in a popular and radical quest for liberation. Suppression from the state became

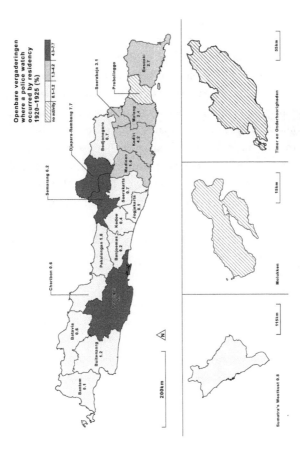

Openbare vergaderingen where a police watch occurred by residency, 1920–1925 (%)

no activity | 0.1–1.2 | 1.3–4.2 | 4.3–7.7

Bantam 0.1
Batavia 0.6
Buitenzorg 1.2
Cheribon 0.6
Pekalongan 1.6
Banjoemas 0.2
Semarang 6.2
Kedoe 0.4
Jogjakarta 0.1
Soerakarta 0.7
Djapara-Rembang 0.7
Bodjonegoro 0.7
Madioen 1.6
Kediri 4.2
Malang 2.4
Soerabaja 3.1
Probolinggo
Besoeki 2.7

Sumatra's Westkust 0.9   115 km
Molukken   15 km
Timor en Onderhoorigheden   55 km

200 km

MAP 12. *Openbare vergaderingen* where a police watch occurred by residency, 1920–25

more frequent, but this frequency of suppression only begat more resistance. As public participation both from the communist and noncommunist organizations became heightened, the repertoires of resistance became varied to include particular concerns of the participants.

This is the backdrop of the revolts of 1926 to 1927. It explains why, despite only happening in few places, the revolts led to the mass arrest of thousands of people and the persecution of thousands of others, both in forms of death sentence and exile to Malaria-ridden Boven Digoel in Papua. This mass arrest explains the popularity of the movement which had become a threat for the sovereignty of Dutch colonial state in the years before the revolts.

The movement was nonparochial in imagining its understanding of solidarity and liberation, despite the existence of party interests pushed by both local and international communist party leadership. It was able to create a universal language of solidarity and emancipation by incorporating the particular needs of ordinary people as it spread and expanded anticolonial resistance and red enlightenment. In turn, anticolonial resistance was not just about ousting Dutch colonialism; it included women's emancipation, reformation of Islam, and the push for universal human rights for equality and justice. Perhaps, it was not the popularity, but rather the universal demands of the movement that became the biggest threat to the state.

The next chapter will investigate further how these universal demands were expressed in the day-to-day production of OVs creating cultures of resistance. By focusing on their qualitative characteristics, we will unearth how this form of revolutionary communication embodies red enlightenment.

OPENBARE VERGADERINGEN AND CULTURES OF RESISTANCE

Isn't this proof that it is not only religions of the prophets that can save the world and mankind, but that ordinary people too can control their own fortune.

—*Sinar Hindia,* April 10, 1924

The history of *openbare vergaderingen* (OVs) in the Indies is an account of participation and socialization in the *pergerakan merah* through the making of common cultures of resistance. Until the first decades of the twentieth century, literacy, assembly, and political associations—despite their emancipatory discourses—remained the privilege of the Eurasian and the *prijaji,* and most lower-class native Indonesians were not invited to participate. But under the *pergerakan merah,* lower-class Indonesians were united, first, as *kromo* people and, later, as proletariats, and they adopted OVs as their vehicles of struggle.

The idea of "ordinary" in the epigraph is key for the movement because for a long time, colonialism, aristocracy, and religion had excluded everyday individuals from the realms of politics and cultures; they were unable to decide and "control their own fortune." "Ordinary" became an important signifier for the struggle against colonialism. While lower-class peasants and workers were central in the economic production of the colonial government and private enterprises, politically they were alienated from the public sphere. To refer to the *kromo* as "*orang biasa*" (ordinary people/commoners) was therefore revolutionary, as it highlighted a story of exclusion from which creativity and hope emerged. The movement created a desired future for the otherwise underrepresented *kromo* people by creating and mobilizing new democratic and participatory cultures of resistance through OVs. This production of OVs as revolutionary communication embodies its campaign of red enlightenment. The beginning of the *kromo* public is the dawn of a long journey toward national revolution that ousted colonialism two decades later.

Previous scholars have noted the importance of rallies and strikes, including OVs, for the *pergerakan merah*. They unearth their roles in the development of radical political parties such as the Communist Party of Indonesia (PKI), Sarekat Islam (SI), and Indische Sociaal-Democratische Vereeniging (ISDV), as well as labor unions and show how embedded *vergaderingen* were in party building at the local level in inland central Java.[1] This chapter expands the importance of this revolutionary communication by analyzing it as a cultural institution. The building of a party does not just consist of a line of commands, a few leaders, political propaganda, and formal events and ceremonies. For a movement to be successfully created, it must transform the mundane aspects of ordinary citizens' lived experiences by creating a new institution of culture. Political life is not superior to, and does not transcend, everyday people. Politics is ordinary because women, children, and rank-and-file members were the active creators of the movement. They did not passively absorb instructions from the Communist International (Comintern) leaders in Russia or from Indonesian leaders. Rather, they collectively created the new cultures of anticolonial resistance.

This chapter turns to the qualitative characteristics of the 800 or so OVs that were mapped in the previous chapter. My findings suggest that these meetings were cultural as much as they were political. Anticolonial politics was produced through the creation of collective culture expressed in forms of entertainment, education, and everyday repertoires of defiance. The "lived politics" of resistance is therefore not only found in palaces, offices, battlegrounds, and town halls, but also in what are traditionally seen as "nonpolitical" settings, including the home, the family, songs, fashion, the arts, and cultures of debate, in which aspirations of social justice, egalitarianism, and emancipation were embedded. These cultures of resistance transformed the *kromo*'s sense of identities, feelings, and consciousness and made politics ordinary. These findings reveal to us that the lasting effect of the *pergerakan merah* was not just the party and its leaders; the cultures of resistance also bore the aspirations of red enlightenment. Even though the movement faced its untimely demise by 1927, it left lasting legacies in terms of lived experiences, identities, and understanding of the world, creating ferment cultures of anticolonial resistance for the National Revolution to come.

A People with Means

The methods by which a community, public, or collective is built within systems that exclude them are key to understanding both the politics of resistance by ordinary people and how colonialism affects one's sense of community and belonging. One such strategy is to examine the communicative means that produce a specific shared identity, ideas, and knowledge of that community. The creation of a collective, which implies a shared consciousness and identity, depends on its means of communication, and these means of communication, in turn, shape the social character of the collective. Resistance is born out of the conditions of alienation and exclusion, and resistance movements usually produce their own means of communication to unite individuals around a similar cause and concern. In the case of resistance in the Indies, the popular communist movement adopted, repurposed, and transformed the existing OVs.

The *pergerakan merah* did not invent OVs. It was a legal term for "public meetings" to be held by political and social associations (see chapter 6). The first recorded OV that was held in the Indies was on December 25, 1912 in Bandung by the leaders of the Indische Party Edward Douwes Dekker, Tjipto Mangoenkoesoemo, and Soewardi Soerjaningrat.[2] Soon, this means of communication was borrowed by SI to organize and mobilize the poor and illiterate *kromo* people. But the emergence of the collective consciousness of *kromo* should be traced in the publication of R. M. Soewardi Soerjaningrat's article written in Dutch titled "Als ik eens Nederlander was" ("If I were a Dutchman") in the newspaper *De Express* on July 19, 1913.

The absence of words often means the absence of consciousness. In the case of Soewardi's article, this absence was resolved through translation. A new consciousness emerged through contacts who spoke different languages and used different systems to understand the world. Written at a time before the Netherlands was about to celebrate its one hundredth anniversary of its monarchy, Soewardi writes

> What a joy, what a pleasure it will be to be able to commemorate a national day of such great significance. I would like, for a moment, to be a Dutchman.... How happy I would be, if later in November the day that I had been waiting for so long came, the day of the celebration of independence. My joy would overflow to see the Dutch flag fluttering happily with a piece of Orange on it. My voice would be hoarse and sing along with the songs "Wilhelmus" and "Wien Neerlands bloed," when the music starts playing. I would be arrogant because of all these statements, I would praise God in the Christian church for all His goodness, I would ask, plead to the high heavens so that Nederland reigns forever, also in this colony, so that it is possible for us to maintain our greatness with power this big one behind us.... I would ... yes I don't know what else to do so, if I were a Dutchman, because I would be able to do anything, I guess.

No, not! If I were Dutch, I wouldn't be capable of anything . . .
I would not want the natives of these countries to participate in that
commemoration.

. . .

But . . . I'm not Dutch. I'm just a brown-skinned son of this tropical
country, a native of this Dutch colony and because of that I'm not going
to protest.[3]

James Siegel explains that "[w]hen one speaks two languages, one of which is
one's first language and the other not, one has two "I"s and one habitually shifts
between them with the possibility always opened of developing different persona
for each."[4] Soewardi's "*ik*" (Dutch for "I") allows him to imagine himself to be
Dutch and plays the persona of a Dutchman. Translating his native reality of
"*saya*" (Malay for "I"/"me") in the world of "*ik*" allowed Soewardi to explore a
different world and to playfully protest colonial conditions.

Soewardi's article is transformative because it gave "*ik*" different subjectivities.
"*Ik*" implies a new possibility for a Dutchman to treat the colonized differently
and a new possibility for a native to be free and powerful like a Dutchman. When
this article was republished in Malay, this transformative effect multiplied as "he
and therefore anyone else could imagine himself [to be Dutch]."[5] It awakened a
different subjectivity in "*saya*." The colonized, passive, and suffering "*saya*" could
then adopt a free and empowered subjectivity.

This phenomenon is a story of finding an enlightened self; however, when this
article was translated, republished, and circulated among the *kromo* it became
a story of finding a community in kind. When *orang ketjil* (lit. small people,
meaning poor people), learning from Soewardi's "*Als ik eens . . .*" started to ask
in Malay "*Djika saya . . .*" (supposing I/what if I/imagine if I), a new world of
possibility was created for them. A *kromo* "*saya*" was awakened and saw that her
emancipation was possible. As this article reached the eyes and ears of its *kromo*
readers, collectively this desire turned into an awakened "we." I became we as
we inheres in I. The identity of small people as an "I," and, together, as a "we"
was infused with new meaning. It is empowered with the ability to change its
condition. *Kromo* as it was mobilized was not an identity to maintain but rather
an identity to overcome and surpass, to no longer be small people. It is not clear
who first came up with the word *kromo*, but it was first used in SI movement to
identify the collectivity of *orang kecil* (lower-class people) who were awakened
and desired for freedom and justice. As the history of *pergerakan merah* shows
us, it is through reading, listening, and gathering that "we" the *kromo* were continuously built. For the collective identity and community of *kromo* to thrive, it
required the production of its own means of communication that represented

the collective needs and goals. Within just a few years, OVs that used to be held by Dutch-educated *prijaji* elites in city centers mushroomed across the Indies in rural and urban towns held by ordinary *kromo* with no formal education.

Openbare Vergaderingen as Entertainment

It was the afternoon of November 4, 1923. Some people were busy decorating the office building in Gendong village, Semarang, where an OV would be held that night. Photos of Vladimir Lenin, Karl Marx, Semaoen, Tan Malaka, Henk Sneevliet, Sun Yat-sen, and Rosa Luxemburg were hung around the center of the room on the bamboo beams that supported the roof. Beneath them, rows of chairs had been neatly set out. Some potted green plants were placed in front of the tables on the podium, which had been adorned with a red cloth. Red flags with hammer and sickle symbols ornamented the walls.

Some people started to gather in front of the building and chatted while waiting for others to arrive. From Djoewana, a neighboring town, the SJS train (Semarang-Djoewana Stoomtram Maatschappij/Tramway Company) heading toward Semarang was full of men and women who were headed for the meeting. The men wore red ties, and the women wore red *kebaja* (Javanese traditional clothing). A group of peasants travelled on foot from the sugar plantations in the northeast of Semarang, each bringing a torch to light their way home later in the night. As the sun set, some 4,000 people from around this coastal city and the neighboring towns and plantations had filled the building. Several government officials and police were seated in the front rows to keep a close watch on the meeting. They were accompanied by journalists from various local newspapers who were eager to report on this popular scene.

Fifteen minutes before the *vergadering* started, children of the Semarang SI school entertained the attendees by singing in a choir, "wearing red pants and white shirts, standing in line singing red songs that attracted the hearts of the viewers."[6] At 6 p.m., *woro* Moenasiah, the chair of the women's branch of Semarang Red SI, opened the meeting by tapping a gavel. She explained the agenda of the meeting, and as she invited the people not to stay silent "like the *blangkon* people" (*blangkon* is a Javanese hat worn by the aristocrats), the attendees laughed. She then invited Mr. Jasin to speak about the exile of Semaoen. As the night wore on, different speakers took turns leading the discussion, from Mr. Soemantri on the banning of *vergaderingen*, to *woro* Djoeinah on capitalist exploitation, Mr. Nawawi Arief on Islam, and *woro* Soepijah on women. Several attendees from the audience occasionally raised their hands to ask questions so challenging that the rest, wanting to hear more, called out loudly, "Debate, debate, debate!" At times, when a speaker talked about the people's suffering under colonialism, the attendees shouted in support; when another speaker gave hopeful remarks about the promise of communism, the audience members simultaneously clapped their hands and shouted, "long live communism!" At 9:25 p.m., chairperson *woro* Moenasiah tapped the gavel again, signaling the end of the meeting. Together, everyone stood up and sang "Marianna" and "Internationale" before heading home.[7]

This reconstruction of a typical scene of an OV, as reported in *Sinar Hindia,* provides a vivid example of how *vergaderingen* had become a common aspect of the people's mundane lives, and how, in this sort of setting, a shared identity and a collective community was built by members and sympathizers of the *pergerakan merah.*

While OVs regularly included political discussions, it was first and foremost a form of entertainment. At a time when electricity, radio, and television were nonexistent, *kromo* people gathered in *vergaderingen* to discuss matters that related to their own suffering after long days of work in plantations, factories, workshops, and homes. They allowed the *kromo* people to release stress and to express and discuss the difficulties they faced in the workplace. Under colonialism, the attendees' lower social status and lack of education led them to accept

FIGURE 5. Female members of the Communist Party of Indonesia (PKI) and Sarekat Rajat (SR). Sitting to the left of the hammer and sickle sign is *woro* Moenasiah.
Source: The Marx-Engels Forum in Berlin, Germany, taken on September 10, 2019. Photo courtesy of the author.

their *kromo* fate. But these meetings gave them a sense of hope and a feeling of belonging, and as such they became avenues for the *kromo*'s entertainment. It was not entertainment in terms of false hope and escapism. Rather it was one that quenched their thirst for freedom from their sufferings. Through the gatherings, for the first time the *kromo* mattered, and for the first time they could speak out to decide their fate. When they turned suffering into hope, they empowered themselves. But when they turned this hope into a collective goal, they created a revolutionary cause. This is the soul of the *kromo*'s entertainment, through which red enlightenment penetrated remote areas.

Vergaderingen, while seemingly simple, became complex creative outlets that had both emotional and intellectual appeal. Strategies to alleviate their suffering through communism were prominent, but they were conveyed by comedy, music, and beauty of all types. The meetings were livened by communist songs and decorative adornments, and communist ideas would often be conveyed through art and fashion. This resistance was not ignited by propaganda or war; it was produced in the realms of feelings and sensations. Resistance changed people's consciousness in taken-for-granted areas like language, relationships, friendship, and art. The very act of creation became political.[8]

Songs

As an important cultural expression of the *pergerakan merah*, songs reflected a global cultural exchange of sources of resistance.[9] During the meetings, people learned to memorize and perform different communist songs, which had been institutionalized as a part of the movement. Like the *vergadering* described earlier, meetings usually began and ended with all the attendees standing and singing "red songs," which became an important part of the *vergadering* agenda and was replicated whenever a *vergadering* was held. Taking part in the movement meant participating in a meeting, which exposed participants to these songs. They came to know by heart songs like "Internationale," "Darah ra'jat" (People's blood; other spellings: Darah rakjat, Darah rajat), "Sair kemerdikaan" (Verses of liberation), "Bendera merah" (Red flag), "Perlawanan" (Resistance), "Barisan moeda" (Young troop), "Enam djam bekerja" (Six working hours), "Meiviering" (May Day celebration), "Proletar" (Proletariat), "Socialisme" (Socialism), "Hidjo-hidjo" (The green), "Roode garde" (Red guard), and "Marianna" (see the lyrics of some of these songs in the appendix).

These songs came from various places: some were composed by Indonesians themselves, but some were translated and adapted from foreign songs. For example, "L'Internationale," a socialist song that was originally written in French, was translated into Malay by the author of "Als ik eens Nederlander was" R. M.

Soewardi Soerjaningrat, also known as Ki Hadjar Dewantara, during his exile in the Netherlands in 1913, a few years before the communist movement emerged in the Indies. Soon after that, the song was adopted as one of the communist movement's anthems. While "Internationale" was a translation from its Dutch version, "Enam djam bekerdja" was invented by railway workers during a strike demanding a decrease in their working hours.[10] This process of translation, adaptation, and invention shows that the movement's cultural expressions, such as songs, were products of transnational cultural contacts as well as political conflicts.

"Marianna"
Saja Marianna, Proletar.
Namakoe telah tersiar.
Saja memakai topi merah.
Tandanja kaoem merdika.
Saja anaknja kaoem rendah.
Djikalau kita merdika
ia 'kan djadi soeamikoe,
jang berani mempihak akoe.
Mari, Marianna!
Pimpinlah kita, toeloeng doenia
dari penindas, bikinlah kita merdika!!

Hai, toekang besi jang bekerdja
dalam boemi ta' liat tjahja
dan kau, jang kerdja di laoetan,
kau tani, penggarap sawah!
Toean kamoe, lintah daratan
printah kau pertjaja Allah,
Sebab ia hidoep di Soerga
Tapi kau hidoep di neraka.
Mari, Marianna!
Pimpinlah kita, toeloeng doenia
dari penindas, bikin merdikalah kita.

Roemahkoe, o, kaoem proletar
Telah lama 'mat terbongkar.
Kau akan makan sama sama,
bila datang persamaan.
Saja reboet wadjib sama
boeat laki dan perempoean.
Demikianlah kita dirikan

Negeri damai, kema'moeran.
Mari, Marianna!
Pimpinlah kita, toeloeng doenia
dari penindas, bikinlah kita merdika!!

"Marianna"

I am Marianna, the Proletar.
My name has been known.
I wear a red hat.
The sign of free people.
I am the child of poor people.
When we are free,
I will take as my husband he
Who dares to be on my side.
Come on, Marianna!
Lead us, help the world
From the oppressors, make us free!!

Hi, blacksmith who works
inside the earth without light
and you, who works in the sea,
you peasant, who works in the paddy field!
Your lord, a leech
commands you to believe in Allah [God],
because he lives in heaven
but you live in hell.
Come on, Marianna!
Lead us, help the world
From the oppressors, make us free!!

My house, o, the proletariat
has long been destroyed.
You will eat together,
when there comes equality.
My struggle should be the same
for both men and women.
That's how we build
A peace, prosperous nation.
Come on, Marianna!
Lead us, help the world
From the oppressors, make us free!!

The movement's cosmopolitan character can be understood further by analyzing the lyrical content of the songs. "Marianna" is a good example. This song depicts an emancipated woman who was hailed as a leader in a movement against the oppressors. "Marianna" was a character adopted from "Marianne," a symbol of the French revolution.[11] She is a female allegory of liberty and reason who represents the republic. The adoption of a female character was meant to personify liberty as an ordinary woman and not as an ancient icon like the respective Roman god and goddess Mercury and Minerva. By the 1790s in France, Marianne, along with red caps, the tricolor flag, and the slogan "Liberty, Equality, Fraternity" had become the standard repertoires of opposition and contestation.[12] More than one song about Marianne existed in different languages, but the Indonesian adoption seems to be translated from a Flemish Dutch song "Marianne," which began with the words: "I am Marianne, proletarians!" with a refrain: "When the hour of vengeance sounds, my spouse will be the name who marches most bravely at my side!" This was then answered by the choir: "Forward, Marianne, guide us, deliver society."[13] The song is saturated with communist terms and language—for example, "Proletar," "free people," "poor people," "blacksmith," "peasants," and "struggle," which introduced and familiarized the common people who participated in the movement with its language.

By drawing inspiration from the French revolutionary character and indigenizing her through the mix of communist words and Islamic vocabulary, Marianna became a symbol for women who joined the *pergerakan merah* in the Indies. Through Marianna, women were invited to be critical of religion, particularly how Islam had been used by capitalists to support the exploitation of workers and oppression of women. As the song puts it, "Your lord, a leech/commands you to believe in Allah [God],/because he lives in heaven/but you live in hell." During a time when the society was still dominated and dictated by the common Javanese and Islamic tradition in which women were seen as the property of their fathers or their husbands, the bold reference to Islam was emancipating. Through the song, the women expressed the desire to be treated as equal to men, to be able to have the autonomy to choose a partner who shared the same emancipatory hopes, and to be a leader in the common struggle against capitalist oppressors. Through this song, the red enlightenment's repertoires of emancipatory language and politics were introduced to women and other members of the movement and became everyday entertainment.

The translation and adaptation of Marianna as a symbol of the communist movement in the Indies further show how emancipatory ideas of freedom and equality became universalized as different local communities struggled for them. One way to look at the context of resistance is through fashion. In one skit before a *vergadering*, an Indonesian playing Marianna took the stage. While "wearing a *koepiah* and a red *kebaja*, letting loose her long hair with a red flag on her left hand and a sword on her right hand, she sang [Marianna] with very loud voice."[14] The *koepiah*, a Muslim traditional hat, and the red *kebaja*, traditional Javanese clothing, are an important modification from the French Marianne, who is often depicted by wearing the Phrygian cap, which signifies women as champions of liberty. This was the cap that was "placed upon the heads of slaves when they were given their freedom" in Rome.[15] For the Indonesian peasants, *koepiah* and *kebaja* were the symbols of their indigenous identity as a Muslim and Javanese. This adaptation was not a form of foreign Western culture that directionally persuaded indigenous people; it was a story of a creative engagement of local citizens who were immersed in cultural aspirations and inspirations from previous global revolutions. It is a story of finding their place in the world.

Traditional Arts

The birth of the politics of resistance through the cultural mixing of the old and the new was also manifested in the transformation of traditional arts. The people of Java had a long, evolving history of traditional arts, such as *wajang* (puppet show), *loedroek* (drama comedy), *ketoprak* (theater), and Javanese songs.[16] During the *pergerakan merah*, revolutionary ideas of resistance were also presented to the public through these traditional art forms. For example, in a *vergadering* to celebrate Labor Day on May 1, 1925, a live-action classical Javanese theatrical performance *wajang orang* (lit. human *wajang*) dramatized the story of Kadiroen that was discussed in chapter 2.[17] First published as short stories in a feuilleton in *Sinar Hindia* in 1920, the socialist novel *Kadiroen* chronicles a native local government official who, after witnessing a communist speech abandoned his career to join the communist movement. An educated man from the rank of the *prijaji*, Kadiroen signifies the movement's campaign to entice the nationalist spirit of the educated natives to struggle for the *kromo* class. This was not an atypical career trajectory during the *pergerakan merah*; many people from the *prijaji* class, like Darsono, similarly joined the movement. *Woro* Moenasiah—a traditional stall (*waroeng*) keeper—was described in a Dutch report as "hung with gold ornaments."[18] One of the movement's campaigns called for the "educated natives" (*prijaji*) to join the struggle for the *kromo* cause; the movement was mobilized through education and enlightenment, as *woro* Djoeinah called for in chapter 2. Taking multimedia forms from a series of short stories in a newspaper, a novel, to a *wajang* orang performance, the movement used available communicative means—both local and transnational, to promote its campaigns.

Other traditional arts were also used to disseminate communist ideas. Barbara Hatley documents this in the use of *ketoprak*.[19] *Sinar Hindia* reports from 1920 to 1924 also listed the use of *loedroek*—another type of Javanese theatrical performance—and *pantoen Sunda*—Sundanese oral performance that integrates narrative and song.[20] The emergence of new political ideas from the blending of longstanding traditions and novel technologies was not unique to the period of communist movement; the adaptation of old forms by the emerging cultures of resistance parallels the ways in which Islam originally spread across Java between the thirteenth and sixteenth century. Islam arrived and was quickly disseminated with the help of traditional art forms, notably *wajang* and Javanese songs.[21]

New Rituals

The *pergerakan merah* also gave birth to new rituals, leading to the creation of a new common identity and a new collective community. The new rituals can be seen most vividly in the structure of OVs. As we can see in the scene described at the start of this chapter, *vergaderingen* became highly ritualistic. They began and ended with the attendees singing together communist songs, usually "Internationale," "Darah rajat," and "Marianna." The meetings were also led by a *voorzitter* (chairperson) and several speakers, such as shown in a photo of a *vergadering* held by SI in the open air in 1916.[22] The *voorzitter* sitting in the center of the photo, next to the podium, was responsible for introducing the speakers and for facilitating the discussion. He had to strike a gavel to signal the beginning and the end of the meeting. Next to him (as in this photo) was a secretary who was responsible for writing a summary of the meeting, which would later be published in a communist newspaper. The meeting would follow this ritual even if it was held in a private house and attended by only a few people. This set of rituals made *vergaderingen* develop into an important cultural institution.

Other rituals central to the movement included holidays. Under Dutch colonialism, holidays that were celebrated by the population in the Indies were either religious holidays, such as the Islamic Eid al Fitr, Eid al Adha, and the birth of the prophet Muhammad, or political or government holidays, such as the birthday of the Dutch queen Wilhelmina. With the emergence of the communist movement, two dates were celebrated as important for the communists. The first was May 1, which was Labor Day, and November 17, which celebrated the Russian Revolution of 1917. On these holidays, *vergaderingen* were held as celebrations of the proletariats. Offices of communist-leaning parties and unions, as well as their newspapers, were closed. The decorations and merriment in the holiday parties resembled the description of the scene from the opening of this chapter, but it often also included children and adults marching around villages wearing red uniforms and singing communist songs.

They also celebrated famous communist leaders. In the same way that the prophets Jesus and Muhammad were seen as holding a special place in Islam, international communist leaders such as Vladimir Lenin, Karl Marx, Henk Sneevliet, Sun Yat-sen, and Rosa Luxemburg became role models in the fight for liberation. It was not uncommon to see large pictures of these leaders hanging on the walls in people's houses.[23] Sold at a small affordable fraction of fl.0.50 to fl.2 (Dutch guilders; the daily wage of Batik Javanese laborers in Semarang in 1924 ranged from fifteen to seventy cents), these photos made communist leaders a regular presence in the lives of Indonesian peasants, workers, and coolies alike.[24]

Openbare Vergaderingen as Educational Institutions

The process of becoming a culture of resistance was also demonstrated in OVs as educational institutions. They were the spaces in which a new modern educational institution was developed for the first time as a collective project. The Sekolah SI "Islamic Union School" (later Sekolah Rajat, "People's School"), as well as the Sekolah Perempuan "Women's School" were products of common culture, a project created by rank-and-file members—men and women alike. Unlike traditional kinds of weaponry, this movement took the education of the people's minds as a key part of its mission and its ultimate success. This education was not one-directional propaganda; instead, the appeal to human senses of critical thinking and education was motivated by the idea that everyone had a shared ability to think, and hence, the right to receive education. These missions embodied red enlightenment.

Before the introduction of a legal category of OV in the Indies, gatherings usually took place as family meetings or in traditional rituals. The structure of the meeting was not egalitarian; it was usually led by the head of the family or the head of the tribe, who were usually men, and the discussions covered religious issues or upcoming family rituals such as wedding, birth, and death. OVs introduced a different structure of meetings wherein women and regular individuals (not religious leaders, men, or tribal leaders) chaired the meetings. With this, OVs also introduced a new culture of debates, allowing conversations to veer away from family dynamics and toward political dialogue.

In this way, OVs are more like the salon culture in Europe than traditional gatherings in the Indies. According to James van Horn Melton, the salon embodied important characteristics of the Enlightenment in the eighteenth century and developed in line with the rise of print culture. Likewise, the conversation in the salon was not exclusively oral; it also generated and circulated the written worlds. Communication was reciprocal and egalitarian, participants were heterogenous, and debate, discussion, and public judgment were encouraged.[25] Women were seen as "civilizing agents" rather than merely as "nature" (they were often contrasted with the idea of men as "culture"). These contours of sociability helped participants in the salon to transcend locally embedded identities and to imagine themselves as part of a larger political community.[26] Despite being adopted over a century after, OVs embodied these features as an enlightenment institution.

OVs were closely integrated with efforts to educate the people, and their discussions included a great deal of critical content. For the audience, which consisted mostly of the *kromo* who had not received formal education in schools and could not read or write, the meetings became an avenue to learn about the political, economic, and social matters that affected their lives. A newspaper, such as *Sinar Hindia*, the organ of the PKI, was often read aloud by one literate person in a meeting so it could be discussed and debated. *Vergaderingen* also became a venue to report foreign affairs, historical accounts, and local issues. These meetings provided a fulcrum between oral and written literacies.

A testament that OVs were an educational institution can be traced in the study by W. M. F. Mansvelt that documented one thousand communist inmates of Digoel prison camp in 1928 following the abortive communist-led revolts in 1926 to 1927. He found that

> at a time when official statistics showed Javanese literacy to be under 6 per cent, his sample was over 75 per cent literate. However, what is interesting is that none had gone to an institution of higher learning and only 2.4 per cent had received middle school training (only one graduated). The vast majority were from the lowest educational institutions in which they were unable to obtain sufficient training to place them in professional or technical fields where prestige and security were to be found.[27]

These internees reflected the population of the movement that consisted of educated *kromo*. Mansvelt helps clarify that the *kromo* became literate not through formal education but through other avenues including the communist cultural and educational institutions such as OVs.

Another characteristic of OVs as an educational institution was the culture of debate. A published invitation was usually accompanied with the phrase "*debat vrij*" (*debat* [Malay] = debate, and *vrij* [Dutch] = free).

> My fellow men and women of all nations and religions.
> Come! Come!
> Debate *vrij*, question *vrij*, talk *vrij*.[28]

Notice the repetitive use of the Dutch word *vrij*, which highlights the welcoming character of the debate. For lower-class people whose lives continuously included suppression and exploitation, the word *vrij*, especially as applied within the realm of debate and discussion, became an invitation to enlightenment. In one *vergadering* led by *woro* Moenasiah, one male peasant named Suprapto raised his hand and asked: "In Islam, men could practice polygamy. What do you think?"[29] This question was not directly relevant to communism; issues that were related to everyday lives were also discussed in the meetings. Anybody could address questions that could usually only be answered by people of higher status. *Vergaderingen* transformed this hierarchy in the traditional Islamic education and allowed Islamic matters to be discussed in open debates, thereby ensuring participatory parity. *Woro* Moenasiah's full response is unknown, but the report included her saying in Javanese—a more informal and intimate language for these people, "*yen durung jelas kena takon maneh*" (if it's not clear yet, please ask again), and Suprapto said "*poeoeoeoen*" (that's iiiiiiiiit). The fact that *woro* Moenasiah invited Suprapto to elaborate on a taboo topic demonstrates the openness of the exchange. More importantly, the code-switching to Javanese, the mother tongue of these people, in the setting of a *vergadering*, in which Malay was normally used, indicated that *woro* Moenasiah wanted to avoid challenging Suprapto with her answer. Suprapto, by lengthening his word "*poeoeoeoen*," sounded relieved and not confronted. These communicative characteristics of OVs demonstrate the defining qualities of the social movement of this period: education, debate, and popular and democratic participation, which became the means for people's organization. These qualities became the new emerging cultures of resistance in the history of the Indies.

Another institution that came out of this movement through OVs was the People's Schools.[30] The Communist Party in the Indies paid serious attention in training their cadres. By 1926, 8,000 PKI members and 100,850 SR members had completed a ten-month course on communism (see chapter 3).[31] However, the education for children that the movement provided included more than just communist training. To ensure that the young generation gained proper basic education, the public meetings of the *pergerakan merah* in 1920 expressed two key concerns: 1) that children could not get a basic modern education, which meant to read, write, learn math, and learn Dutch; and 2) that many of them were forced to work, often on plantations, to help their parents to fulfill the family's economic needs.

While the Ethical Policy, which was installed at the beginning of 1900s, included education as one of the Dutch colonial government's main goals to improve the welfare of the native population, two decades later the majority of them still could not earn basic education. In response to the upcoming Congress on Native Education in May 1922, *Sinar Hindia* noted that, despite the fact that the children themselves demanded education, the government could not fulfill it because of the lack of teachers, high tuition, and poverty. "Every year, tens, nay, hundreds of children cried because their dream to go to school was not fulfilled. Not enough seats." Both Hollandsch-Inlandsche School (HIS, Dutch School for Natives) and the second-class school, usually located in villages or mountains whose students were mostly the children of sugar *kromo*, imposed high tuition for the *kromo* class. As sugar plantation labor, these parents could not afford the tuition of fl.0.50. They were so poor that even cloth became another hindrance for these children to go to school. "Oftentimes they [students] could not go to school because the [only] cloth was used by their parents to go to the market."[32] With these problems, it is not a surprise that most native lower-class children did not have access to government education. As the table shows, only between 14.7 and 29.8 percent of the total population of children, depending on the age group, had access to government-provided education. In 1928, between 70.2 and 85.3 percent of native children remained entirely uneducated. As a result, many of these children helped their parents by working to earn additional income for the family. In response to this lack of access to existing Western education, native *prijaji* created modern schools infused with local ethos, such as Taman Siswa and Muhammadiyah.[33]

People's Schools were created in response to the lack of access to education for the *kromo*. Together with Taman Siswa and Muhammadiyah, they were labeled "wild schools" to differentiate them from schools that were formally funded and held by the colonial government (see chapter 6). People's Schools were founded through *vergaderingen* and the communists in the *vergaderingen* agreed that People's Schools had to be different from the Dutch schools.[34] The

TABLE 7.   Number of native children with access to government schools in 1928

YEARS IN SCHOOL	ACCESS	% OF NATIVE POPULATION	NO ACCESS	% OF NATIVE POPULATION
3 Years	1,431,429	29.8	3,372,200	70.2
5 Years	1,623,745	29.3	6,386,200	79.7
7 Years	1,647,761	14.7	9,568,578	85.3

Source: Perhimpoenan Peladjar-Peladjar Indonesia, *Menentang Wilde-Scholen Ordonnantie* [Resisting the Wild-School Ordinance], January 1933, the Royal Netherlands Institute of Southeast Asian and Caribbean Studies (KITLV), Leiden, the Netherlands. Data on the number of native children with access to government-provided education can also be collected from *Volkstelling 1930*, a voluminous report by the Dutch on their first comprehensive attempts to provide a colonial census. Dutch East Indies, *Volkstelling 1930: Census of 1930 in Netherlands India* (Batavia: Department van Landbouw, Nijverheid en Handel, 1933).

schools' motto explains this motivation: "the domination of the capitalists stands on education based on the logic of capital. To resist this: people's freedom can only be earned through education based on the people [Malay: *kerakjatan*]."[35] The schools taught Marxist concepts such as capitalism, labor, and value, in addition to fundamental skills like Dutch, math, sports, reading, and writing, as well as cooking and sewing. The creation of the institution was decidedly ordinary; these schools were built and developed by members of the movement, and private houses were adopted to be spaces for learning and studying where SI schools took place. For example, E. J. (Betsy) Brouwer, Sneevliet's wife, who stayed in the Indies with their two twin sons Pim and Pam when Sneevliet was exiled in 1918, was a teacher and helped decorate schools in Semarang and donated books for the students.[36] To fund the schools, as William Frederick notes, *kromo* students in both rural and urban areas were given the flexibility to pay the fee in forms of rice, eggs, fowl, or other foodstuffs.[37] Up until 1925, People's Schools mushroomed across the Indies (see map in chapter 3).

Tan Malaka was the first person to initiate the creation of SI schools.[38] In one photo, he poses with the children and other teachers, many of them women.[39] The children wore uniforms of white shirts and red pants. Those standing and sitting in the first two front rows held red flags, some of which bore the hammer and sickle.

People's School became a place where lower-class natives were socialized with new self-understanding and consciousness that transcended their immediate locality, giving rise to a new idea of community that was global and cosmopolitan. These were achieved through a shared experience of communal learning. Students from People's Schools occasionally attended *vergaderingen* and helped teach the illiterate adults. These children, according to Malaka, were "wise and brave to give a speech" and "*vergadering* functioned in order (with rules) and with firm heart (from both the speakers and the listeners)."[40] These schools trained children with leadership skills producing "*tukang pidato proletaar*" (skilled proletariat speakers) and "*ahli pena*" (pen experts).[41] These meetings became an avenue to introduce revolutionary vocabulary and ideas, such as "*revolutionair*" (revolutionary) "*kapitalisten*" (capitalists), "*kapitalisme*" (capitalism), and "*rintangan*" (determination). Students would share what they learned in schools with the *vergadering* attendees, including adults who were never able to attend school. Outside of Java, schools also became an important institution for the development of the movement in the Outer Islands. In West Sumatra, communism began in the Islamic school Sumatra Thawalib, starting a new political tradition in which Islam and communism were mixed.[42] The exiled sailor Djamaloedin Tamin, which will be discussed in chapter 7, was one of its teachers. It is not clear if these communist schools decreased the number of child labor in the plantation, but these children likely continued to work part-time after school to earn income for the family.

In December 1920, the women's branch of SI in Semarang, led by *woro* Djoeinah, wrote a letter to the city council requesting a school special for women, published in *Sinar Hindia*.[43] This demand was further discussed in an article titled "The Movement of SI Women Semarang."[44] It seems unlikely that the council honored this request because a few years later in June 1922, in the midst of the burgeoning educational program led by the communist movement, Djoeinah launched and taught "Snel School Salatiga" (Quick School of Salatiga) for adult women. In an article she lamented that many women were still prohibited from attending by their husbands, so she advised the men to support the women's goal to become educated.[45] Another leader, *woro* Roessilah Hadisoebroto from Ngandjoek, also suggested that an informal school for women was to be established in a "public kitchen" so women could educate themselves on recent political issues while preparing meals for their families.[46]

Openbare Vergaderingen as Cultures of Defiance

OVs gave birth to various cultures of resistance. They appealed to the human instinct to "negate." In the case of this movement, this entailed rebelling against the compartmentalization of identity, the domination of religion, and the extension of repression in family relations. Between 1918 and the beginning of 1923, *vergaderingen* functioned primarily as entertainment; however, after the banning of *vergaderingen* between May and September following the 1923 strike, the number of *vergaderingen* swelled and meetings became demonstrations of open defiance, not just against colonial repression of the freedom of speech but also against the government's attempt to compartmentalize political groups based on identity.

In line with the popular tone of the meetings, people of different classes, genders, and religions were urged to join the debate and the movement. A common phrase, placed at the end of an invitation distributed through newspapers and leaflets, stated: "all men and women from all nations and religions are invited."[47]

Partai Kommunist India
Branch Semarang

ATTENTION! ATTENTION!
Big Openbare-Vergadering
On February 3, 1924

On Sunday at 9 AM.
In the office of ISLAMIC-UNION Gendong.
What will be discussed:
The fate of our brothers who were arrested
Because of bombing.
This is a really important matter!
That is why we hope the people in Semarang,
men and women, come together en masse in this *Vergadering*!
Check and tune into this bombing matter.
Make time to come!

And take your friends to come along!
THE BOARD[48]

The call for men and women from all nations and religions was an important campaign, as the movement was facing antagonistic reactions that sought to dichotomize Islam and communism. The idea that Islam and communism were not compatible was the basis of much anticommunist propaganda. The Dutch government also created the compartmentalization of different political groups into Islam, communism, and nationalism to make the three seem as though they conflicted with each other.[49]

The compatibility of Islam and communism was indeed one of the most debated topics in *vergaderingen*. At the time, people were divided between those who saw communism as a godless idea that posed a threat to Islam and those who felt that the practices of Islam could be strengthened through communism. These debates resulted in efforts to reinterpret both Islam and communism. The Muslims in the *pergerakan merah* were confronted with explaining how ideas of liberation, resistance, and justice—all key communist ideas—also had roots in Islam. In the newspapers of this period, discussions on this reinterpretation of Islam on issues of social justice were common. Arabic verses of the Quran were juxtaposed with communist concepts to explain how in fact they were compatible. It is important, as Takashi Shiraishi warns us, to not use the misleading label "Islamic communism" as if there were several kinds of Islam (e.g., Islamic imperialism or Islamic capitalism).[50] The communists did not argue for "Islamic communism"; rather they campaigned for the synthesis of the two, "Kommunisma and Islamisma" (Communism and Islamism).[51] The conjunction "and" here is key.

During this period, some practitioners interpreted Islam by defining it in terms of universal concern with social justice. One of the famous red hajjes was Haji Misbach.[52] An article called "Islam and the Movement" that Misbach wrote emphasized the universal idea of humans able to live peacefully together, to take care of each other, and to resist capitalist exploitation.

> Men and women live in this world to live peacefully together. That is God's command. To live peacefully together, we have to follow God's commands. If somebody chooses to follow God's commands while not caring for other people's needs, he has instead committed sins and instead has followed the devils' will. Devils are our enemies. In this century, they work for capitalism. Capitalism is a lust for destroying our belief in God. Communism taught us to resist capitalism, because that idea is already covered in Islam. I explain this as a Muslim and a communist.[53]

Here the terminology of Islam, like "God's commands," "sins," "devils," and "belief," is juxtaposed with key terms of communism, like "live peacefully together," "capitalism," and "resist." Islam was reinterpreted in terms of resistance and social justice, while communism was expressed through the language of religion.[54]

The compatibility of religion and communism was not just demonstrated in writing or speeches but also in the collective organizations that mobilized under communism. Indeed, since its emergence, religious institutions had been the foundation of the communist movement. In West Sumatra, Muslim leaders such as H. Datoek Batoeah turned their Islamic school into a communist organization. Leaders in Timor also melded religious and political ideals, whereas Christian Pandij spread communism through Christian beliefs in churches.[55] Despite the purported rift between Islamism and communism, members of the movement defied such "politics by exclusion." While this example shows us how difference was politically produced, it also shows how the movement universalized itself by accommodating these differences. Eventually, openness to any religions, sexes, and nationalities was crucial in making the communist movement and its OV a democratic and popular space.

The cultures of resistance promoted through *vergaderingen* also emancipated the most mundane and ordinary group: the family. Among communist women, for example, the critical thinking that became the culture of *vergaderingen* helped them to redefine their relationship with their husbands. This was hinted at several times in speeches, and is detailed in a story titled "The flow of time or a girl who suffers," in which a woman from the audience in an OV argues[56]:

> My family! Now I no longer have a husband. Because when in his workplace there was a strike, my husband refused to join the strike. Those

who did not know me would think that it was me who hindered him from joining the strike. At the time when my husband came home and said he didn't join the strike, I asked for a divorce. I am not lying! I think it'd be better to be a divorcee than having a husband with a long tail [a coward]!

It is not clear if this quote reflects an actual occurrence, but the fact that the writer included this episode shows the common expectation for women to be liberated both in marriage and in the movement. While communist women often asked for a divorce from husbands who did not want to join the movement, the reverse was also true. In a discussion of *reactie* (reaction against communism), women often divorced their communist husbands because they feared anticommunist harassment. Divorce was common in Java among women of diverse social backgrounds. In 1930, the divorce rate in Java was 5 percent versus 1.5 percent in the Netherlands thanks to the Islamic tradition of repudiation (*talak*). According to the tradition, when a man spoke the *talak* formula three times, divorce occurred without any requirements, including alimony. It was the Christian-influenced Dutch officials who condemned divorce and, by the early 1900s, wanted to develop marriage morality and regulations based on Western norms. They wished to improve marital conditions and discourage divorce through Western education and influence of native elites.[57] For communist women, however, divorce was a means to voice and claim liberty and freedom. The critical culture that was encouraged by OVs also began transforming the social practices among the people.

How did the women explain the daring move in their domestic life? According to Djoeinah, women were oppressed in both domestic life and public life, in their past and their present, and by family as well as by strangers (in this case, the capitalists). In front of about a thousand people attending an OV on September 26, 1920, *woro* Djoeinah gave a speech on women's oppression:

It is clear that the crime committed by money people [capitalists] will chop down the world. . . . this will not only suffer men. Women will also be destroyed due to the high price of basic goods. In this period, the wage that workers receive is not enough. So wives are forced to work as well. . . . Money people are happier to employ female workers because their bodies [labor] are cheaper than men. This is a threat to the unity of men and women. . . . Many wives also sell their body [for sex] for money to the capitalists. This has united workers to struggle against the capitalists. But at the same time the government also tightens our movement by controlling our meetings.[58]

Here Djoeinah explains that the politicization of women's bodies within colonial capitalism occurred in two ways. First, by exploiting their bodies as commodities, that is as labor power to sell in exchange for wages; second, through the emergence of more widespread prostitution, that is, sexual activity as a specific form of labor. Elsewhere, Djoeinah explained that just as the capitalist mode of production oppressed women by turning their bodies into commodities, women were equally oppressed by the men who controlled women in "the grasps of their hands." The problem of women's condition as the property of men and parents that had been the character of "*adat kolot*" (old tradition) was given serious attention by the movement. Another writer said that women were still "tied and hindered by disastrous men and parents who were still holding to old traditions."[59] It was important for women to join the movement and for women and men to unite in the struggle, as the struggle against capitalism meant the struggle against the commodification of women's bodies.[60]

Another act of defiance was the transformation in naming practices. Parents began to give their children communist names instead of Javanese names. However, adults also adopted names with communist roots. In Javanese society, it is not uncommon for people to replace the name they received shortly after birth to a new name when they reach adulthood. A Javanese individual usually selects a name that seems most fitting to his circumstances—perhaps one that refers to a new job or a social position. Still, adults could adopt a new name in the hope to change their fortune.[61] During the communist movement, many of them adopted names rooted in communism: "Samirasa" (equality, also "Sama Rasa"), "Sovjeto" (from Soviet), "Mintasama" (another word for equality), "Hardjoproletar" (Hardjo the *proletar*), "Siti Sowjetika" (Siti is an Arabic name and it is combined with *Sowjetika* from Soviet) "Indier" (the Indies person), and "Si Kromo" (the lower-class). After the 1926 revolt, the PKI was banned, and hundreds of thousands of its members were arrested and exiled to Digoel, some 2,000 miles away from Java in the Malaria-ridden land of Papua. One photo shows some of the PKI members who had just arrived in Digoel. Four of them sitting on the chairs were women. The boy squatting in the middle is "Sovjetto," named after "Soviet."[62] Many rank-and-file members adopted these names to show that equality had been absent from their lives and how they longed for equality in their fight for justice.

Red Enlightenment as Culture of Resistance

The history of OVs tells how red enlightenment became commonplace through the development of its rhetoric in entertainment, education, and the family. It was created by common people in mundane settings and affected the cultural and political structure of their lives. Communist politics was expressed in the whole lived experiences of the people and was manifested in diverse cultures of resistance. In names, songs, family relations, religion, expressions of art and the mind, aspirations of resistance that were rooted in red enlightenment seeped into the structure of feeling and the senses of the people. It appealed to the human senses of beauty and enjoyment, of the ability to think, and the desire to negate and resist. Politics was created in day-to-day settings, and in turn the ordinary became political.

Through the process of translation, negotiation, circulation, and transnational co-production, contacts between cultures and between the new and the old created new common cultures of resistance. Alongside the anticolonial struggle, women's rights, children's needs, and the interests of people of other nationalities were also forwarded. Their dreams and hopes began to be materialized in the very act of resistance. These cultures of resistance were not just local and personal; they were also global, universal, and human.

"THE WAR OF PENS AND WORDS"

A communist movement without a newspaper is like a person fighting a tiger with no hands.

—"Pemberiantahoe. Penting!" ["An announcement. Urgent!"], *Api*, July 30, 1925

The growth of the native vernacular press and political organizations in the first two decades of the twentieth century resulted in the rise and fall of the "revolutionary press" (*pers revolutionair*) in Indonesia.[1] This press was an outgrowth of the earlier vernacular press, which, along with political parties and unions, had become a voice of the colonized people throughout the Dutch East Indies. As the quote above suggests, the *pergerakan merah* vehemently chose the revolutionary press as one of their main weapons of struggle, alongside *openbare vergaderingen* (public meetings, OVs) and People's Schools. The role of the revolutionary press in a broader project of red enlightenment in the colony was even more significant.

During this period, the explicit goals of the *pergerakan merah* were not limited to ridding the Indies of colonial capitalism; they also included elevating a new intellectual movement. On January 2, 1922, an article in the *Sinar Hindia* (later *Api*, henceforth *SH/Api*) newspaper captured the passion for "the new way of thinking" quite vividly:

> For most people here in the Indies, it has been difficult to embrace new *wetenschap* [Dutch: knowledge], mostly because they do not possess *the new way of thinking* [emphasis in the original]. Their *geest* [Dutch: spirit] is still old-fashioned, and belief in superstition still plays an important part in their way of thinking. Remnants of a more religious era, an ancient time, still reign in the minds of the people in the Indies . . . To promote the movement and the new *wetenschap*, we have to try as

hard as we can to pry them away from their misguided thinking by promoting our propaganda through brochures.[2]

The image of the new was a preoccupation of the movement writ large. Here,
"new knowledge" and "the new way of thinking," expressed in a seamless juxta-
position of Dutch and Melayu, were considered key to freeing the people's spirit
from the misguidance of religious dogmas, mysticism, and superstition.

Earlier scholars have discussed the role and development of the press in shaping national awakening in the Indies. The desire for the new was manifested in
the idea of the nation as a form of collective community. Since the rise of modern
political parties in 1908, the idea of an independent Indonesian nation emerged
as the uniting idea among the colonized subjects. As Ahmat Adam suggests, the
rise of this consciousness emerged through the vernacular press, that is the press
produced and circulated in Melayu by and for the natives.[3] Benedict Anderson
explains through his notion of "print capitalism" that a nation is "an imagined
community" in which the colonized organized themselves and formed a community by creating a sense of belonging that reached beyond their regional, physical, ethnic, and linguistic differences. This imagining was possible because of the
press and other commodities of print capitalism that facilitated the creation of
nations.[4] Scholars have also recognized the importance of the vernacular press
in aiding the rise of parties as modern forms of political associations that were
indispensable to the birth of the new period of *pergerakan*—"an age in motion"
or "World-in-Motion."[5] Nevertheless, it is important to emphasize that, during
the *pergerakan merah*, Indonesian nation as an identity and a movement was not
yet formed fully.

This chapter unearths that during this "age in motion," alongside existing
nationalist movements, the *pergerakan merah* imagined its collective community
beyond the Indies and created demands beyond national rights insisting instead
for humanity and universal freedom and justice for all. Framing these enlightenment practices primarily as part of a national struggle subsumes the former
as a parochial project of the latter. In reality, the *pergerakan merah* was a project of enlightenment that mobilized anticolonial struggles using the language,
aspirations, and goals of universal emancipation. Studies on the revolutionary
press in Turkey, the United States, France, and Iran have highlighted the role
of the press in enabling public opinion, public discussion, and public criticism.
The enlightenment practices of the press have also helped to organize collective
actions, often leading to the overthrow of a monarch, regime, or colonial rule.[6]
Investigating how enlightenment practices developed in the Indies can help us
understand how those cultural practices, while unique, were at the same time
facilitated through the production of books, newspapers, and magazines. Most

important, in the case of the *pergerakan merah*, enlightenment practices were born out of political struggles against colonialism.

This chapter rescues the tradition of the *pers revolutionair* by examining the production and development of the newspaper *SH/Api* from its conception in May 1918 until its closure in April 1926 as an exemplary case of the revolutionary press. This chapter demonstrates that the revolutionary press emerged out of the tradition of the vernacular press. Through an investigation of the paper's production and distribution practices, as well as its textual aspects, this chapter reveals how *SH/Api* embodied the anticolonial national struggle and became one voice for a project of enlightenment in the colony.

The Evolution of *Sinar Hindia*: A Brief History

By the time the newspaper *Sinar Hindia* (literally, "light of the Indies") was first published in 1918, the push for enlightenment had already begun. This is apparent in the names of many newspapers that circulated in the first two decades of the twentieth century. Their titles included words such as *sinar* (light), *suara* (voice), and *bergerak* (in motion), demonstrating the relationship between the struggle for enlightenment and the desire to resist colonial rule.

The development of the vernacular press in the Indies began on January 25, 1855, when the Javanese-language weekly *Bromartani* was published by Carel Frederik Winter, a Eurasian born in Yogyakarta whose career included working as a Javanese philologist and a Javanese language translator for the government.[7] The vernacular press continued to grow with the involvement of the Eurasians and the Chinese in the industry, many of whom trained Indonesian natives to become printers and editors. By the end of the nineteenth century, native readership within a small, but more educated, class had grown. When the native vernacular press was finally established, it functioned not like its predecessors, which were committed to supporting commerce; instead, it was explicitly committed to supporting political struggles. This commitment of the press had been inspired by the movement in Chinese communities within the Indies to turn the vernacular newspapers into political organs and create a common identity of Chinese nationalism in support of the Chinese revolution.[8] Between 1903 and 1913 the native vernacular press and political organizations developed in tandem. Tirtoadhisoerjo created Sarekat Prijaji and its organ, the *Medan prijaji*, in 1906 to 1907; Dr. Wahidin Soedirohoesodo, the editor of the periodical *Retnadhoemilah*, created the Boedi Oetomo party in 1908; a Eurasian journalist named Edward Douwes Dekker was in charge of *De expres* when he founded Indische Partij in 1912; and Tjokroaminoto was the editor in chief of *Oetoesan Hindia* when he began leading Sarekat Islam (Islamic Union, SI) in 1913.[9] However, many of these political parties were only able to attract educated elites, many of them coming from the *prijaji* class.

Unlike the earlier generation of native newspapers that were produced by journalists-turned-activists, *Sinar Hindia* was produced by activists belonging to both Semarang-based SI and Partai Komunis Indonesia (Communist Party of Indonesia, PKI). The name changes in the history of *SH/Api* reflect an increased focus on enlightenment in the movement's development of revolutionary press. When several Indonesian communist leaders had begun joining SI and founded its more radical Semarang branch in 1918, they bought the *Sinar Djawa* (light of Java) newspaper and changed the name to *Sinar Hindia*. The new editors, Semaoen, Mas Marco Kartodikromo, and Darsono, no longer imagined the newspaper as only the organ of the people of "Java," but rather of the entire "Indies," as indicated by its masthead, "the newspaper for *kromo* people in the Indies." During the movement, the word *kromo* was typically used when talking about the oppressed and the colonized. In 1924, the newspaper changed its name again, this time to *Api*, and created a new masthead: "the voice of the proletariats of all nationalities and religions." *Api*, meaning "fire," still conveyed a sense of light or enlightenment, but it was also more aggressive, signaling the further radicalization of the movement.

The shift in the newspaper's masthead from the "newspaper for *kromo*" to a paper for "proletariats of all nationalities and religions" was more than a terminological change. It reflected a deeper split in the organization. The organizational split into a "white wing" and a "red wing" resulted from a disagreement about whether Islamism and communism were mutually compatible ideologies. Eventually, the white wing of SI refused to stand together with the communist members, afraid that they would betray their belief in Islam in favor of communism. The red wing of SI responded by claiming not only that Islam and communism were compatible, but that it was in the interests of the Muslims that the communist struggles against Dutch imperialism should succeed. To fight Dutch rule, they suggested that the *pergerakan merah* should not use a religious banner—given that people of other nationalities and religions in the Indies had fought against colonialism—and instead should use nonreligious symbols to unite the diverse anticolonial contingents.

The use of the term "proletariat" to replace the Javanese term "*kromo*" also represented a reframing of the party's membership, which was largely peasant, since the term proletariat is often understood as referring to industrial workers. Most people in the Indies at this time remained in rural areas, and while the movement gained support from urban workers in port cities such as Semarang, Batavia, and Surabaya, it also mobilized the rural lower classes.[10] But why did SI think of

FIGURE 6. *SH/Api* editors before the exile of Semaoen in May 1923. The same editors who produced *SH/Api* also produced the organs of the VSTP, *Si Tetap* and *De Volharding*. This is the room of the VSTP board of directors. Sitting down, from left to right: Semaoen (1), Soedibio (editor in chief of *Si tetap*) (4), Abdoelrachman (propagandist) (5), Soegono (editor in chief of *De volharding*) (3), and Kadarisman (secretary) (2). Attached photos left to right: Zainoedin (committee from Aceh), F. A. Zeijdel (treasurer) (6), Mohamad Ali (treasurer) (7), and Abdoelwahab (committee from Padang) (9).
Source: Hoofdbestuur, *Poesaka V.S.T.P.* (Semarang, Indonesia: Drukkerij VSTP Semarang, 1923), access no. 2.20.61, no. 144, Inventaris van de Collectie Documentatiebureau voor Overzees Recht, 1894–1963 (Inventory of the Collection Documentation Office for Overseas Law, 1894-1963), Nationaal Archief (NA), the Hague.

its membership as proletariat? Its use of the term can be read in two ways: (1) it affirmed the organization's affiliation with the existing international communist movement and (2) it unified the diverse membership under a new proletarian identity. As the social movement with the largest following in 1924, the latter fact points to the inclusive nature of the movement, accommodating people of other nationalities—not just the natives—who supported the anticolonial cause.[11] The use of proletariat as a common, everyday word resonates with Anderson's idea of "World-in-Motion universalism" from which new representational imaginings of revolutionary movements in the Dutch East Indies and abroad were born.[12] During this period, new political terms such as *vergadering* (meeting), *kongres* (congress), *mogok* (strike), *klassenstrijd* (class struggle), and *revolutie* (revolution) appeared and began to be used and interpreted not as "autochthonous," but rather "universal" categories, enabling the people to imagine themselves as a part of a new global universe—"Workers of the world, unite!"[13]

Following the internal split within SI, the red wing of SI changed its name to Sarekat Rajat (People's Union, SR) to distance itself from Islam. Between 1924 and 1926, red SI, SR, and the PKI were essentially operated by the same people—those who fell under the banner of communism. The change of name to *Api* and SR symbolized a revolutionary point in the history of native political associations in the Indies in that they were able to spread the movement across diverse constituents and move toward a more inclusive democratic community.

Understanding the Claim "Revolutionary"

The publishers, editors, and journalists behind the *SH/Api* newspapers self-identified as "the revolutionary press" (*pers revolutionair*) with the expressed purpose of representing the voices and interests of the *kromo* people. A close reading of their understanding of the revolutionary press, however, uncovers a more important development in the vernacular press. In making the case for the revolutionary press, they also effectively revolutionized the press by developing journalistic practices that embodied red enlightenment ideas.

In an April 9, 1924 article titled "The Freedom of the Voice of the Press. Sowing the Seeds of Hatred?," Synthema—the pen name of Soemantri, who took over as editor of *Sinar Hindia* when Semaoen was banished from the Indies in 1923—defined the goal of "the revolutionary press of Indonesia" as "the effort to seek an improvement for a better fortune and liberation for the Kromo class."[14] Soemantri wrote this in response to the accusation made by D. A. Rinkes, an adviser for native affairs in the Indies who would later establish Balai Poestaka to promote the publication of reading materials and the foundations of libraries for the indigenous population.[15] To Rinkes, the revolutionary press spread the "seeds of hatred" between classes. Using the language of the colonial *haatzaai artikelen* (hatred-sowing articles),[16] Rinkes suggested that the government release a new regulation requiring all newspaper publishers to pay insurance in the amount of 5,000 *roepijah*.[17] According to Soemantri, as Rinkes threatened,

> If that press has a revolutionary direction and does not listen to the warning of the head of the government up to three times, then the money will not be returned and the newspaper will be banned from being published.[18]

Rinkes's attitude against the revolutionary press was shared among mainstream newspapers in the Indies, including the Dutch broadsheet published in Surabaya *Soerabaijasch handelsblad* (Surabaya Commercial Paper; *Soer Hbld*), which represented the voices of the conservative European population in the Indies as well as the interests of sugar plantations.[19] On April 14, 1924, *De Sumatra Post* reviewed an article from *Soer Hbld* under the title "Against Propagating Communism." According to the *Soer Hbld* Batavia editor who wrote the article, the freedom of press was "abused" by "an intensely malignant extremely nasty indigenous press acting against the West," which often received help from European publications, providing camouflage to support communist actions and leaders. Agreeing with this article, the *De Sumatra Post* argues that both preventive and repressive measures need to be taken—the removal of communist public servants from government offices, making hate speech punishable, taking firm control over both native and European presses—and expresses its support to "assist the government in the quiet fulfillment of its task" and "to promote the introduction of the aforementioned measures."[20]

This conflict reflected a broader power struggle between the government and mainstream newspapers, on one side, and the movement, on the other, in assigning different meanings to the revolutionary press. Whereas Rinkes saw the revolutionary press, especially the one associated with SI, as troublesome, Soemantri saw it as necessary to realize an accountable government.[21] In the same article, Soemantri asked, "How could stealing the government's money [corruption] that has put people at a disservice happen?" and he answered, "It is because we do not have the right to control and criticize. And if we dare to express criticism to keep watch over crimes not to happen, these writings were considered *sowing hatred* [emphasis in the original]."[22] Even at this time, the revolutionary journalists had seen their role as gaining control by criticizing the activities of the government. Soemantri's article is revealing in two ways. First, he saw the need for a newspaper to represent the voices of the *kromo* people, which was clearly absent in the mainstream papers. Second, in his demand to create an alternative space for the *kromo*, Soemantri offered a new concept of the role of the press as a means of keeping the government in check. While he sought to make a case for the revolutionary press, Soemantri offered revolutionizing ideas about the press in general making the case for newspapers' function for "critical publicity."[23]

Journalists as Defenders

The agenda to turn *SH/Api* into a revolutionary newspaper was consciously intended to differentiate it from mainstream newspapers. In discussing the role and job of a journalist, Darsono, under the pen name D. A. S., wrote an article arguing for a differentiation between "journalists of rice/bread" ("*journalis roti of nasi*") and "journalists as defenders" ("*journalis pembela*").[24] This distinction was triggered by an article written by a peranakan Chinese journalist with the pen name Koetoe Bolspik (Bolshevik flea), who, while observing the mushrooming of new young journalists, also lamented their qualities. He called them "*tjingtjau*," which here is used metaphorically to refer to "mushrooms [growing wild] in the rainy season."[25] The word was used to describe the mushrooming of journalists who took the job for money and who produced writings that "lowered their status" as journalists.

In his article, Darsono made a comparison between journalists of rice or bread and journalists as defenders. He wrote:

> Indeed, in the world of journalism there are two categories, journalists of bread or rice and journalists of "the defenders."
>
> Of course the destiny of these two journalists is different. Journalists of bread, if they happen to find themselves in trouble because of their writings, it is just an accidental matter. This is because they already take side with the people or the class who have power over bread (bread here means living necessities), so everything for them is better and more liberating than it is for the journalists of the defenders.
>
> The journalists of the defenders in general consist of two groups: the defenders of the workers and peasants, and the defenders of our land and nation [nationalist]. These two groups of journalists—their mind and action—are not interested in money, but are interested in their knowledge and their belief. But, the nations who have power over the world, the oppressors, they really hate these revolutionary journalists, who defend those who are oppressed and humiliated. Not only do they hate them, but they also create a law to protect themselves so they do not lose their power. . . .
>
> In our group there are many who have become the victims of their writings. But, our enemies' purpose to oppress our journalists has not succeeded. Instead it leads the proletariats to think that if the front row gets oppressed, the back row will replace them by rolling up their sleeves and sharpening their pens. One [writer] down . . . one [writer] up![26]

The contrast is clearly made between journalists who did not actively involve themselves in the movement, seeing the badge of journalism as simply a profession, and those who actively took risks and launched an active defense for the people. The contrast is also made between the revolutionary and the "reactionary," "capitalist," or "white" press. The former were journalists who appealed to knowledge. They engaged in dangerous work, were always at risk of arrest, and were viewed by the government and police as troublemakers who sowed the seeds of hatred. The latter were journalists by profession who worked simply for the money (for bread/rice). These journalists, instead of joining the movement, "slept at home and then wrote all of [their] dreams in their newspapers." Respectful journalists "will not sensationalize an event, the news of which will not be shortened or added. They will just take the essence, so people understand because newspapers should be educative, propagandistic, defensive, and helping for our class."[27]

The contrast that Darsono highlights points to a larger problem of the press in the Indies. Colonial capitalism seemed to have led to the creation of journalism as a career, leaving behind its political potential as bearer of reason and truth. By questioning the integrity, duty, and responsibility of the journalists of his time, Darsono demonstrated that being revolutionary journalists meant more than just defending the *kromo* class; in their reporting, journalists also had to seek truth and reason, and to disdain sensationalism.

In addition, Darsono's likening the duty of journalists with that of soldiers "rolling up their sleeves" points to how the revolutionary press recruited writers and editors from the ranks of the proletariat, inviting them to join the struggle by "sharpening their pens" to write. As on the battlefield, Darsono wrote, "one writer down, one writer up."[28] This militaristic metaphor also resonates with what the media scholar Joshua Atkinson calls the "decentralization" process allowing for "multiple people to play important roles simultaneously . . . so that as circumstances develop different people can step up and fill in any voids."[29] This is an important mechanism in light of the persecution of *SH*'s writers and editors. Darsono's writing strikingly demonstrates the adoption of reading and writing as weapons of struggle. This shift in the movement reflected its broader commitment to the promotion of a project of red enlightenment.

Darsono was at the forefront in the call for a more accountable journalism with integrity and dedication to reason and truth. Darsono was an editor for *SH/Api* and made his way to journalism via the movement, another key characteristic that distinguished *SH/Api* from the vernacular press. Whereas the previous generation of *pergerakan* newspapers was produced by "journalists-turned-professional" leaders, *SH/Api* was led by propagandists-turned-journalists.[30] This postwar generation of propagandists such as Semaoen (b. 1899) and Darsono (b. 1897) were also younger than the previous *pergerakan* leaders, such as Mas Marco (b. 1890) and Tjokroaminoto (b. 1882).[31] Mas Marco began his career in the *pergerakan* as an apprentice in *Medan prijaji* and was trained by the father of native press, Tirtoadhisoerjo. When he led the editorial board of *SH/Api* with Semaoen and Darsono, it was likely because he was a senior and well experienced journalist of the *pergerakan* at the time, able to train the propagandists-turned-journalists that dominated the editorial and printing staff of *SH/Api*. Since Semaoen and Darsono grew up during a time when the language of resistance and anticolonial consciousness had already become part of everyday life, the writings of Semaoen's generation were more direct, fierce, and strong—similar to the language in their speeches, rallies, and strikes.

As the "organic intellectuals" of the movement, they were working-class intellectuals, creating the revolutionary press as part of a struggle against colonialism.[32] Even if some of them were native intellectuals from the educated higher class *prijaji* background, like Darsono, they identified themselves with the *kromo* class and joined their cause. Soemantri took over the editorial position of *Sinar Hindia* when Semaoen was banished from the Indies in 1923. Other editors included Samsi and Djoeinah, who was *SH/Api*'s first woman editor. *Woro* Djoeinah was the leader of the women's branch of SR in Salatiga—a mountainous town near Semarang.

Journalists of this period played dual roles as both writers and movement leaders. Semaoen and Darsono, like other editorial members, were leaders in the *pergerakan merah*. The correspondents of the newspapers, who were also members or sympathizers of the movement, came from all over Java, West Sumatra, and several Outer Islands which had the greatest numbers of Red SI, SR, and PKI followers.[33] For the next eight years, the newspapers recruited revolutionary journalists from the rank-and-file membership of these parties and unions, and those journalists led strikes and public meetings.

SH/Api was by no means the only newspaper that saw the revolutionary mission as a part of a larger project of enlightenment by the movement. The

TABLE 8.   List of *SH/*Api editors

YEAR	EDITOR IN CHIEF
1918–23	Semaoen
	Soemantri (1923)
1924	Soekindar
	Soemantri
	Soebakat
1925	Soebakat
	Heroemoeljono
1926	Soewitowignjo
	A Mangoensamarata
	W Kamsir

emergence of the revolutionary newspapers was in fact a general trend in other cities across the Indies. Like *SH/Api*, other newspapers that shared its characteristics and openly voiced the interests of the proletariats or lower-class people included *Neratja* (Batavia), *Oetoesan Hindia* (Surabaya), *Sri djojobojo* (Kediri), *Benih-Mardeka* (Medan), *Hindia-Sepakat* (Sibolga), *Heroe-Tjokro* (Kediri), *Matahari* (Bandung), *Padjadjaran* (Bandung), *Soeara-Ra'jat* (Semarang), *Tjokrowolo* (Kediri), *Ma'loemat* (Fort de Kok), *Ra'jat bergerak* (Solo), *Halilintar* (Pontianak), *Panggoegah* (Yogyakarta), *Soeropati* (Sukabumi), *Djago! Djago!* (West Sumatra), *Proletar* (Batavia), *Soeara Tambang* (West Sumatra), *Pemandangan Islam* (West Sumatra), *Njala* (Batavia), *Doenia achirat* (West Sumatra), and *Pandoe merah* (the Netherlands).[34] Their editors were also leaders in the communist-leaning movements in their respective regions.

The idea of "revolutionary" carried a specific meaning in this movement. It criticized the way in which knowledge and journalistic products were produced within colonial capitalism, and practiced an alternative form of knowledge production that was tied directly to the larger project of anticolonial struggle. This enlightenment project during the movement was vibrant and came from below. This contrasts with the notion of "revolutionary" decades later, when the word became entangled with the national revolution against the Dutch that ended in 1949. Since then, Soekarno—the first president of Indonesia—used it to motivate a nation-building project. The bottom-up, emergent, creative, and vibrant energy that the idea of "revolutionary" carried in the early communist period was replaced with the top-down language of bureaucracy and formal party politics during Soekarno's time. After the putsch in 1965, the word "revolutionary" was given a bad reputation under the Suharto regime and would be replaced with the spirit of "nationalism."[35] The tradition of the revolutionary press in the *pergerakan merah*, however, appealed to the interests of the lower-class public, and in the process became an institution that produced enlightenment ideas and practices.

The New from the Old

When the newspaper *Sinar Djawa* was founded at the end of 1913 (its name was changed to *Sinar Hindia* in 1918), it was purchased from a Chinese firm, Hoang Thaij and Co. At the time, the prime objectives of *Sinar Djawa* were still commerce and trade; thus, aside from subscriptions, advertisements were still the main source of revenue for its operation.[36] When *Sinar Djawa* became *Sinar Hindia*, however, the editors sought to change the production system by changing its financial sources.[37]

SH/Api was among the first newspapers in the Indies to strongly criticize capitalist encroachment in newspaper production, which it claimed compromised
the quality of reporting.[38] In an article on the cost of subscription, the editors
explained that they were against the support of big capitalists who "sucked the
sweat and the energy of the workers in the Indies."[39] As "a weapon of workers who
are weak and oppressed," they invited the *kromo* to advertise their businesses in
SH/Api, and asked the readers to buy goods and services from merchants who
placed advertisements. "By so doing, Ra'jat [lower-class people] could strengthen
their business and organ!"[40] *SH/Api*'s attitude against capitalist commerce and
trade consolidated when, on May 16, 1924, the SR of Semarang bought the printing company N. V. Sinar Djawa for 3,200 guilders.[41] The production of *SH/Api*
was placed solely in the hands of this radical party. The criticism against capitalist
newspapers uncovered a new concept among the producers: that it was important
to continue the support for the businesses of lower-class people while fighting the
domination of the colonial government and more powerful capitalist industries.

However, because advertising from lower-class business could not generate
much income, *SH/Api* relied on subscriptions as its main source of revenue and
to expand its readership. There was no complete annual report on the exact number of subscribers over the years, but at the end of 1924, *Api* reported that it
earned a total of 36,999 guilders, of which 26,438 guilders came from subscriptions and 5,004.93 guilders came from advertisements. Based on that year's subscription yield divided by the price of a single copy at 0.10 guilders and a monthly
subscription cost at 1.70 guilders, the newspaper had 1,101 to 1,295 subscribers.[42]
This marked an increase from 720 in 1918 to 1,126 a year later.[43] Normally a
newspaper during the period would be considered popular if it reached 2,000
subscriptions; however, despite the seemingly lower subscription, newspapers
that were tied to "The Sarikat-Islam . . . are the best known."[44] *SH/Api*'s large
readership can be detected not from the number of subscriptions, but from the
distribution mechanism that helped spread the newspapers. As the paper repeatedly announced, there were five duties of *SH/Api* subscribers:

1. After reading, pass *Api* on to those who could not afford subscriptions.
2. Send articles on events based on facts to the editorial staff.
3. Help find new subscribers.
4. Help find new advertising clients.
5. Pay subscription cost ON TIME![45]

The first "duty" itself ensured that the number of actual readers was much higher
than the number of subscribers. Further, the number of people familiar with
the newspaper and its content would increase further, as the newspapers were
often read aloud to groups in public meetings. Just like the propagandists in the

movement, existing subscribers were responsible for helping find new subscribers and new advertising clients, so the expansion of the readership would eventually expand the movement itself.

The fact that *SH/Api* lasted for eight years attests to the viability of subscriptions to finance its production. Still, the paper struggled, and these struggles
affected the working environment for its staff. Until the end of its time in print,
SH/Api was produced on used printing equipment "from the 1700s" that was
still "manually operated by hand rather than powered with steam or electricity
like other big printing houses." The machine was old and often broken, imped-
ing the publication for a day or two. Once, "[we] just began printing the second glass yielding only 1100 pages, the printing machine broke and the letters

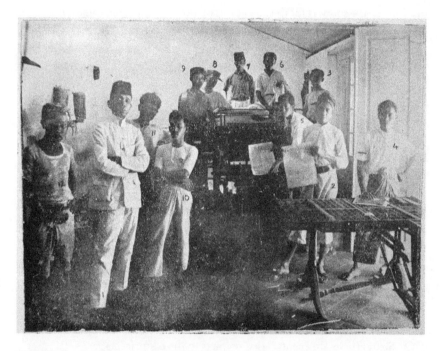

FIGURE 7. *SH/Api* printing staff. *SH/Api* and VSTP printing staff in the printing office, left to right: Oesin (12), Partondo (head of printing and administration) (1), Bibit (11), Doeldjalil (10), Markimin (9), Djaspan (8), No-ong (7), Koming
(6), Satiman (5), Drachman (head of printing) (3), Kasman (head of typesetter)
(2), and Kaslan (4). Partondo was the first person to translate Karl Marx's *Communist Manifesto* into Malay (see advertisement in *Sinar Hindia* on July 2, 1923).
Source: Hoofdbestuur, *Poesaka V.S.T.P.* [The heritage of the VSTP] (Semarang: Drukkerij
VSTP Semarang, 1923), access no. 2.20.61, no. 144, Inventaris van de Collectie Documentatiebureau voor Overzees Recht, 1894–1963, Nationaal Archief (NA), the Hague.

fell sprawling on the floor."[46] The machine had been purchased from a Chinese printer, but the fact that it was still manually operated indicated this machine may have been passed from one printer to another.

This photo indicates the press must have been built in the 1800s. Cast iron began to be used in the early 1800s; prior to that, hand presses had changed very little and were typically made out of wood.[47] The press in the photo appears to be a cylinder press. As industrial printing machines, cylinder presses allowed for a mass production of prints and hence became popular for newspapers. Though first invented in the 1780s, they only became popular after having been perfected in 1840s.[48] However, cylinder presses were meant to be operated either by steam or, later, electricity. *SH/Api* printers had to operate it manually, possibly because of the cost of continuous access to electricity.

Besides the doddery printing equipment, the editors also complained about their small and narrow printing office, pointing out that "many typesetters fell sick due to the small and unhealthy working environment."[49] The photo shows alphabet metal type cases in the foreground, which prove that they indeed used movable types—the kinds of types that were used by modern printing presses first invented by Johannes Gutenberg, explaining why letters fell sprawling on the floor.[50] The use of movable types meant that there was a finite amount of types that could be assembled into words and lines of text. This technological constraint can be seen in the sample invitation in the section "Public Reading as Political Participation" in which the equals signs were used in place of hyphens. The technology and the means of production demonstrates that the daily production of the papers required intensive manual labor. It is likely that the twelve printers depicted in the photo would share the narrow room as they divided the tasks to produce and print the daily *SH/Api*.

Compare this condition with one of the largest printing houses in the Indies at the time, the government-owned Balai Poestaka. A book published in 1929 depicts the government's printing office with its large modern printing machine in a spacious room.[51] This picture contrasts with the cramped, manual setup of the antigovernment revolutionary presses at *SH/Api* and *Si tetap/De volharding*. Even the amount of the papers shown in both pictures proves this difference: the government operated with abundant resources, which is apparent from the stacks of paper in the mid-to-front right of the photograph; only two sheets of paper are shown in the whole room in the *SH/Api/Si tetap/De volharding* photograph.

It is not clear how the lack of funding kept these newspapers going; however, it seems that the Union for Railway and Tram Workers' (VSTP) people were motivated by reasons that were unrelated to salary. Some of them did not just work as printers or typesetters; they also helped contribute articles. Partondo, the VSTP's head of printing, second left in the photo, was also the translator of the first Indonesian version of Karl Marx's *Communist Manifesto*, which was first published serially in *Soeara Ra'jat*.[52] Some of the editorial staff, like Semaoen, also worked for *SH/Api* while working full-time to publish *Si tetap* and *De volharding*.

From late 1925 to early 1926, the government began harassing people who read *Api* by searching their homes to confiscate reading materials. As a result, many subscribers did not want to pay for subscriptions. *Api* eventually shuttered in April 1926 due to a lack of funding. Sustaining a revolutionary newspaper would always be a struggle. Yet, the reverse was also true: With enough popular support and a free legal environment, a revolutionary newspaper—one supported by subscription and donation—could flourish.

A Culture of Criticism and Debate

On February 28, 1922, a slogan at the bottom page read "Sinar Hindia akan memberi *penerangan* [emphasis added] oentoek memboeka pikiran baroe" ("*Sinar Hindia* will give *enlightenment* to begin a new thought"), which illustrated one of *SH/Api*'s projects as a revolutionary newspaper: to build a new system of enlightened thinking.[53] *SH/Api*'s conscious characterization of itself as an institution producing new lines of thought seems to rightly perceive "media as epistemology," as institutions that produce knowledge and meaning.[54] Viewing the need to produce new knowledge through what Mas Marco called "*berperangan penna dan perkataan*" ("the war of pens and words") as a part of an anticolonial struggle is telling, especially given the prevalence of violent resistance through warfare for two centuries prior. As an epistemological institution, *SH/Api* sought to revolutionize through enlightenment. In the full words of Mas Marco in Malay,

> *Soenggoeh soedah terlaloe pajah kami berperang penna mereboet kemenoesiaan kita. . . . Dari itoe saudara-saudara! tahankanlah berperangan penna dan perkataan.*
>
> Indeed, it is too hard for us to wage a pen war to seize our humanity. . . . That's why, comrades! continue the war of pens and words.[55]

The war of pens and words was waged to seize the people's *kemanoesiaan* (humanity). Describing their endeavor in terms of "war" gives a clue to the danger these writers and organizers faced. Regulations and laws were later created by the colonial government to punish the action of penning and publishing (chapter 6). In a time of total suppression, clandestine sailors such as Djamaloedin Tamin endured risks to continue to wage a pen war for the movement to defend enlightenment's ideals of humanity (chapter 7).

However, enlightenment voices were not unified; they were a cacophony of contrasts, debates, and conflicts, and the newspaper saw criticism and debate as necessary to allow such conflicts to be voiced and witnessed by the general public. These were expressed through not just analyses and news reports, but feuilleton, caricatures, poems, innuendos, and public gossip. Darsono versus Tjokroaminoto was a notable case of the latter. In 1912 to 1913, when the SI party was first founded, its leader, Tjokroaminoto, was so popular and well respected by the ordinary masses that he was seen as a messiah. This myth was deconstructed a few years later in 1920, when Darsono, a young radical leader from SI's Semarang branch, criticized Tjokroaminoto for a possible corruption case. The article in which Darsono explained his criticism was published in three parts. The first part was a lengthy justification for why a "criticism" was needed for the movement and the collective organization:

> We have to express this criticism because we believe that with a criticism all mistakes and fraudulence can be fixed and then prevented [so as] not to affect our movement and association. . . . A criticism, as we perceive it, is like a soap that can clean all dirt. . . . A criticism is an effective medicine that can cure all diseases that can lure the association and other order.[56]

The argument expressed by Darsono highlighted the need for a culture of criticism on the internal workings of the leaders. This culture of self-criticism was deemed important for the collective good. Like "soap" or "medicine," criticism could fix, clean, and cure problems.[57] Darsono explains in detail the accusations against Tjokroaminoto (Tjokro) over the case of a fraudulent use of party money in the second part of the article. Darsono's profane treatment of Tjokro's image demands clarification as a part of Tjokro's duty as a leader. He pointed out that the people had collective rights to know the truth, an idea that was foreign even for the *pergerakan*. It expressed a pursuit for openness and publicity disavowing seclusion and secrecy.

In line with the hierarchical nature of Javanese culture, Darsono's criticism did not just create a controversy; it also divided the people. People wrote letters and opinion articles accusing Darsono of gossip and claiming that it was very un-Javanese to criticize Tjokro, who was believed to be immune from sins and mistakes. Some welcomed this criticism as a healthy tradition for the movement. Tjokro himself wrote a response article in *Neratja*, later reprinted in *SH/Api*, brushing off all the accusations as untrue and refusing to address them.[58] Heated debates followed, with articles such as "Darsono ngamoek" ("Darsono Runs Amok"), "Kritiknja kommunist Darsono pada pimpinan C.S.I. berekor" ("Communist Darsono's Criticism against Long-Tailed CSI Leaders"), and "Kritik Darsono!" ("Darsono's Criticism!"). Soon enough, the words *menjokro* (to *tjokro*) and *tjokroisme* (*cokro*ism), which alluded to Tjokro's unresolved case, were used to describe an act of "corruption" whenever corruption cases were reported. The criticism against Tjokro expressed in a possible act of slander and gossip demonstrates how public debates and dialogues did not always come in a form of "sober, clearly argued debate."[59] In fact, this altercation represented a more serious turn for the movement.

The divide was exacerbated by the news of Vladimir Lenin's statement on the national and colonial questions that were uncompromising on pan-Islamism, expressed in the Comintern's (Communist International) second congress (July–August 1920). There, Lenin said:

> It is necessary to struggle against the pan-Islamism and the pan-Asian movement and similar currents of opinion which attempt to combine the struggle for liberation from European and American imperialism with a strengthening of Turkish and Japanese imperialism and of the nobility, the large landowners, the clergy, etc.[60]

The news of Lenin's hostile position vis-à-vis Islam shook the *pergerakan* in the Indies. Progovernment newspapers took this opportunity to attack the communists, claiming that they were passively following imported communist ideas. On February 12, 1921, the PKI released a statement to clarify the relation between Islam and communism, but this was not enough to avoid the split that followed. Articles such as "Kommunisma dan Islamisma" ("Communism and Islamism") and "Kommunisme dan igama" ("Communism and Religion") appeared in the subsequent months, beginning a long debate on the compatibility of communism and Islam. At this point, SI followers were divided between those who saw communism with suspicion and those who believed that communism strengthened the Muslim cause for anticolonial liberation. Anticommunist SI members saw communism as creating trouble for the unity and strength of the SI party. Not long after, the debate on communism and religion morphed into a discussion on *partijdiscipline* (party discipline), suggesting that SI should separate from the communist-leaning members and leaders. Though the actual break between red SI and white SI did not happen until 1923, the intense sentiments against and for communism were first triggered by Darsono's attack on Tjokroaminoto and were aggravated by Comintern's policy on pan-Islamism. *SH/Api* editors had introduced a new culture of criticism and debate.

Public Reading as Political Participation

Although *SH/Api* was initiated by an educated segment of both *prijaji* and *kromo* classes, it was by no means only distributed among the educated Indonesians. Given that by 1920 only between 4 and 12 percent of indigenous men and between

0.1 and 1.8 percent of indigenous women in Java could read and write, how did *SH/Api* manage to play a significant role in a popular movement?[61] The key was in the practices of public reading. *SH/Api* was not primarily consumed in private, but rather in public events such as OVs. Public reading was not the invention of the *pergerakan merah*—it was already a widespread practice at the time. However, it was during the time of the *pergerakan merah* that public reading developed into a form of political participation, integrated with political meetings (*vergaderingen*) and rallies held by the parties and communist-affiliated unions. It was not uncommon for a meeting to require its attendees to bring a particular issue of a newspaper. The meeting's agenda would be to discuss an article in the paper deemed relevant for the attendees. See a sample invitation to a meeting asking attendants to bring *Soeara Ra'jat* newspaper 5 and 6.

Partai Kommunist India
Afdeeling Semarang

Nanti pada hari REBO malam KEMIS tanggal
26–27 MAART 1924, moelai djam 7 malam,
Mengadakan Leden vergadering, bertempat di kan=
toor V.S.T.P., oentoek meneroeskan membitjarakan
voorstel=voorstel Congres.
Saudara=saudara lid harap datang dengan mem=
Bawa bewijs van lidmaatschapnja dan Soeara=Ra'jat
No. 5 dan 6.
HET BESTUUR.

Communist Party of India
Semarang Branch

On WEDNESDAY, the night before THURSDAY
26–27 MARCH 1924, starting at 7 PM,
a Leden vergadering [member gathering] will be held in of=

fice of V.S.T.P. to discuss
motion=motion Congress [Congress motions].
Fellow=fellow members please come by br=
inging their membership card and Soeara=Ra'jat newspaper
number 5 and 6.
THE BOARD.[62]

The synergy between verbal and written cultures explains the relation between the movement's media and nonmedia communicative networks. In this case, newspapers supplied information to be discussed in the *vergaderingen*—nonmedia

FIGURE 8.   Public reading.
Source: Balai Poestaka, *Bureau voor de Volkslectuur. The Bureau of Popular Literature of Netherlands India. What It Is and What It Does*, 26.

communicative networks—and in return, these interpersonal communication networks supplied new material to be reported and debated in the newspapers, making both the media and interpersonal communications constitutive and indispensable parts of the movement.

During *vergaderingen*, newspapers were usually read aloud to the attendees. News was reported, opinion articles were discussed and debated, and political actions such as strikes, donations, and protests were organized. Pramoedya Ananta Toer argues that the early native vernacular newspapers played a role in shaping political opinions. In this period, newspapers organized the people and facilitated collective actions.[63] The culture of debate was integrated into the tradition of public reading and public meetings, which allowed for greater political participation and democratization of ideas.

Enlightenment as a Practice of Contention

The revolutionary press, emerging out of the tradition of the vernacular press, took a different role in the anticolonial struggle in the Dutch East Indies. Instead of merely creating "imagined communities," the revolutionary newspapers became the handmaiden of "organized communities," alternative *kromo* public spheres that constituted countercultures whose voices otherwise remained absent in colonial Indonesia.[64] The revolutionary energy that was exemplified through the activities of revolutionary press, however, was not limited to national struggles against colonialism. It was dedicated to producing practices and ideas of red enlightenment. As a part of the movement, journalistic roles were redefined, ethical journalistic practices and cultures of criticism and debate were promoted, and public reading was adopted as a form of political participation. Likewise, public debate, conversation, and criticism took place within the movement, placing openness and publicity on the agenda as key features for the struggle. Literacy in turn became an area through which political solidarity was diffused. The movement's vision of a future society that was free from colonial rule was embodied within these practices. This vibrant, spontaneous, and voluntary energy resulted in bottom-up, inclusive, and democratic networks of political parties and unions that constituted the "age in motion."

To think of an enlightenment project as a project of anticolonial and anticapitalist struggle carries with it at least several important implications. First, it promotes the ideas of literacy, education, and the public sphere as themselves forms of "popular contention."[65] Official institutions of democracy, such as the state, often emphasized these ideas through the prism of national interests. Second, by understanding it within the context of the movement's anticolonial struggle against Dutch rule, the project of enlightenment that was mobilized within the movement cannot be read as merely a product of the West; rather, it was the product of a complex amalgamation of tactics and inspirations from other movements across the globe. This is apparent in the communist language the movement adopted, the print technology it used, and the global political discourses within which it situated its own struggle. Third, producing red enlightenment as contentious communicative practices means the very act of communication itself is a risky endeavor that required both courage and creativity. The global circulation of knowledge and revolutionary ideas involved creative camouflaging of reading materials and sailors smuggling them, as the next two chapters show more vividly. The revolutionaries risked jail time, exile, and even death. The subsequent chapters further examine how communication practices and technologies during the *pergerakan merah* became circuits of struggle.

IN THE NAME OF PUBLIC PEACE AND ORDER

Judge: Is it true that you, Ms. Ati, gave a talk in an *openbare vergadering* in Donomoelio on August 6, 1925?

Ms. Ati: Yes.

Judge: What kind of *vergadering* was that and did you get a permission from the local government to hold it?

Ms. Ati: We did.

Judge: During the *vergadering* did you not see if there were police and where they were?

Ms. Ati: I did! But I did not know each one of them and where they were [situated].

Judge: Before you gave a talk, didn't the police warn you already not to talk about something that violates the state's law?

Ms. Ati: No!

Judge: Did you deride government officials saying: "In the past, women chose *prijaji* to marry, [but] now you shouldn't wait for them." Right?

Ms. Ati: Yes!

Judge: What did you mean by that? Who were the *prijaji*? Didn't you know that most *prijaji* refer to government officials?

Ms. Ati: I said *prijaji* were not just government officials, but everybody who receives a high salary and is a white collar ["working soft jobs"] is also a *prijaji*.

Judge: "Even if one kills me, I still will not marry a *prijaji*." Didn't you say that?

Ms. Ati: No!

Judge: How so?

Ms. Ati: After my talk was stopped by a police inspector, it is true that I said: "We are women in motion. Even if [we are] prevented and killed, [we] will not give up, before gaining independence."[1]

On December 25, 1925, in a secret meeting at Prambanan in Central Java, after key leaders had been exiled or fled the Indies, the remaining leadership of the Communist Party of Indonesia (PKI) agreed to launch an armed uprising against the colonial government. It was initially planned for June 1926; however, due to disagreements within the party and the continued repression of Dutch authorities, the rebellion did not occur until much later. The communist revolts broke in two areas in West Java, Banten and—to a lesser extent—Batavia, on November 12, 1926, and in West Sumatra on January 1, 1927.[2] Chaotic and disorganized, the revolts were quickly crushed by the Dutch government and were soon followed with the banning of the *pergerakan merah*.

West Sumatra and Banten became key sites of the revolts after communism transformed the political and social makeup of these areas. Scholars agree that Indonesian communism had "the kind of leadership that was the really novel feature of twentieth century unrest in Indonesia."[3] The movement was led by educated and cosmopolitan party leaders and organizations. Studies on this movement have thus far focused on key leaders, their arrests, party lines and commands, and the repression of this leadership and organization through violent policing, surveillance, execution, and imprisonment. Among the best-known Dutch responses to the revolts took place at Digoel, a concentration camp for political prisoners. Digoel was located in a remote area in Papua and was established within two weeks after the communist revolt in West Sumatra. Thousands of alleged communists and the *pergerakan merah* followers were sent there, and Shiraishi argues that the colonial government enforced peace and order in the postrevolt Indies by establishing Digoel as a threatening "phantom."[4]

The remaining history of the revolts, however, is an enigma. The movement was popular and wide in scope. So why did the revolts only occur in three areas in the entire Indies? Why did only a few people join the rebellion? The revolts were conducted only by a few individuals in a few places. Why were thousands of people arrested? This book so far has shown that there was something novel and sophisticated about the *pergerakan merah* outside its leadership and party structure. The strategy of mobilization and socialization through revolutionary communication embodied their emancipatory aspirations of red enlightenment, which led this period to a unique event of mediation.

When we shift our attention away from the leaders and party commands to ordinary people in ordinary settings, as this chapter shows, we learn that repression and violence from both the government and horizontal organizations had become a part of the daily life of the movement long before the 1926 to 1927 revolts. In fact, throughout 1926 *openbare vergaderingen* (OVs) had ceased to exist, and People's Schools were no longer operating. Likewise, revolutionary newspapers stopped their publication by mid-1926. The popular movement, as the *pergerakan merah* is usually described by scholars, had ceased to exist months before the communist rebellion. Therefore, the revolts did not involve the popular voices of the *pergerakan merah*; most were too afraid to mobilize. However, the Dutch government took the revolts as representative of the movement, which provided a pretext for the government to thwart the campaign entirely. The Dutch response to the revolts was equally novel; for the first time it treated communication practices and technologies as warfare. The liberal enlightenment ideas of free press, assembly and deliberation, and education that the Dutch had once promoted in the Indies and later gave birth to red enlightenment aspirations was replaced with conservatist politics.

Communication technologies and practices have historically been a circuit of struggle, a battleground for power between the state and the people. To fully understand their significance, and the broader significance of the movement, we need to locate this communist project dialectically among the contradictions that are generated by colonial capitalism. We must explore the project that countered it, from the government and other horizontal groups. Communicative practices, including the adaptation of communication technologies and the invention of law and regulation, are primarily projects of struggle, and communicative practices are both conditions and forms of agency from which a new order emerges.

This chapter views communication as a political and cultural "circuit of struggle." It analyzes the regulatory framework that the Dutch colonial state created around communicative means and practices as a response to the *pergerakan merah*, in particular the rationale behind three legal products of the Dutch colonial state: "the Press Banning Ordinance," "the Wild School Ordinance," and the repression of "the Rights of Association and Gathering." The laws were installed by different local residencies during the peak of the popular movement from 1923 to 1925, and by the beginning of 1926 until 1932 they were formalized nationally by the Dutch colonial state across the Indies, creating a systematic repression against communism after the revolts. In examining them, this chapter questions the perversion of "public peace and order" that had often become the state's pretext to create suppressive anti-enlightenment policies. Exploring this trajectory unearths how, long before Digoel, the "phantom" had already been created through restrictions and attacks on the movement's daily practices of revolutionary communication. In this case, "*zaman normal*" (the normal age) in the postrevolt era through to the National Revolution was built through both the phantom world of Digoel and through the suppression of expression and political thoughts and ideas.

From Ethical State to Police State

The legal systems surrounding communicative practices in the Indies were transformed dramatically in the first few decades after the turn of the twentieth century. In an effort to bring about prosperity, progress, and education for the natives, the Dutch Queen Wilhelmina installed a new period of "Ethical Policy" in 1901. The Ethical Policy provided the context in which new media of the period facilitated the emergence and development of the *pergerakan merah* in the Indies. However, even though it was inspired by liberal enlightenment, it was not benevolent. It was a variant of the British "White Man's Burden" and the French "Mission Civilisatrice": By providing modern education and introducing legal rights for the natives, the colonial state heightened its control of the social and cultural aspects of the natives' everyday lives.[5] Despite being accessible at the beginning only to a smaller segment of higher-class natives, the introduction of modern education, legal rights to public gatherings and political associations, and the opportunity to publish newspapers for the indigenous population eventually led to the development of new and more innovative strategies of anticolonial popular resistance. Instead of relying on weapons and warfare, the *pergerakan merah* developed collective actions around new emerging communicative technologies and practices that were accessible to the indigenous populations—schools, public debates, popular journalism, arts, and literature—to organize themselves and resist colonialism.

Before the *pergerakan merah*, much of the regulatory environment was created based on ethical ideas of "freedom and autonomy." Under the adviser for native affairs G. A. J. Hazeu (1906–12) and later D. A. Rinkes (1913–16), as well as A. W. F. Idenburg, the governor-general of the Dutch East Indies from 1909 to 1916, the Ethical Policy became a regulatory system that prioritized transparency and freedom. In 1913, the ethical drive led T. B. Pleyte, the minister of colonial affairs in Holland, to defend the freedom of expression in print and in public meetings and regional meetings. On May 8, 1915, this was realized in Article 111, Regeringsreglement, which recognized the right of association and gathering for all residents in the Dutch East Indies.[6] Volksraad, a people's council in which representatives came from various groups, including European, Chinese, Arab, and other races, was founded a year later.

Soon however, the government wanted to set limits around the meaning of the term "ethical." With the development of native associations, Idenburg argued that they could not tolerate parties that questioned the power of the government.[7] Over time, this "limit" was translated into the need to protect "public peace and order." For Idenburg, freedom needed to be followed with order and bound by rules; "order and power" were more important than Western values of freedom and equality, especially in the context of a colonial society.[8] For the colonial officer B. J. O. Schrieke, "public order" meant maintaining harmony between different groups in the society, which lent a pretext for a repressive legal approach in the Indies.[9] Punishment was required for those who spread hatred against authority. This is not unique of the Indies; "public order" is often used as a measure of the limit for civil liberty and an excuse for policing.[10] Over just less than two decades, the regulatory environment shifted from the ethical ideas of freedom, autonomy, and transparency to the discourses of order, security, and policing. In the late 1930s, after her arrest in Digoel, a communist woman named Soekaesih who fled to the Netherlands penned a brochure titled *Indonesia, een Politiestaat* (*Indonesia, a Police State*) with her colleague G. van Muenster showing that the people of the *pergerakan merah* were aware of this transition to a police state.[11]

Ironically it was during the *pergerakan merah*, which clearly adopted enlightenment aspirations, that the Dutch campaign favoring liberal strains of enlightenment came to an end. Conservative voices heavily questioned and criticized the ethical propounders. In the words of Harry Benda and Ruth McVey, "[b]efore long, however, the pendulum started to swing away from welfare policies, innovation, experimentation and liberalization. . . . Little by little the pre-Ethical conservatism in colonial policy . . . gained the upper hand." The communist revolts completely ended the ethical reformism, and conservatism rose in triumph.[12] This conservatism and its legal, political, and social oppression lasted until the National Revolution a few decades later were all indicative of a standard colonial response.

Like the *pergerakan merah*, which borrowed strategies and language of enlightenment and universal emancipation from other movements, colonial governments also borrowed strategies of repressions from each other. For the Dutch East Indies, the model came from the colonial British Indies. In 1913, when R. M. Soewardi Soerjaningrat released the article "Als ik eens Nederlander was" ("If I Were a Dutchman"), which created chaos in the government and the public, the government realized that it needed to control the press. The minister of colonial affairs, Jan Hendrik de Waal Malefijt, consulted with the minister of foreign affairs, R. de Marees van Swinderen, to determine whether British India's Press Act of 1910 would be viable in the Dutch East Indies. The act required a newspaper to pay a bond that could be retained by the government if it violated regulations. The two ministers received an enthusiastic note from London, noting that "[t]he Act has worked successfully. The tone of the newspaper press in reference to politics has much improved."[13] This administrative strategy would serve as a workaround for the Ethical Policy of 1906, which technically disallowed censorship in the Indies.[14]

Although the administrative control around the press was first rejected by the Dutch colonial government and was not implemented until 1932, the government did employ a repressive law from the British Indies, that is *haatzaai artikelen* (hate speech articles). In the *Wetboek van strafrecht* (*Book of Criminal Law*) Article 24 number 2 and Article 26 on printed matters, attempts to incite hatred and insult in prints were punishable; however, attempts brought about in speeches, meetings, plays, and nonprinted manuscripts could not be addressed by legal sanctions.[15] *Haatzaai artikelen* allowed for punishment of these communications.

Based on Articles 124a of the Indian Penal Code, in 1914, and with little fanfare, the colonial government inserted these new rulings around hate speech to the *Book of Criminal Law*.

> Article 63a: He, who by words, signs or depictions or in any other way gives rise or promotes feelings of hostility, hatred or contempt against the government of the Netherlands or the Netherlands Indies, shall be punished with a penal servitude sentence of five to ten years.
>
> Article 63b: He, who by words, signs or depictions or in any other way gives rise to or promotes feelings of hostility, hatred or contempt against different groups of Dutch nationals or residents of the Netherlands Indies, shall be punished with imprisonment varying between six days and five years.[16]

The punishment for hatred against the government in these articles was detailed as follows: "imprisonment from five to ten years for Europeans, and with forced labor with chain from five to ten years for Indonesians." And, for hatred, insult, and animosity incited toward other residents of the Dutch East Indies, the punishment was "six days to five years for Europeans and forced labor 'without chain' maximum of five years for Indonesians."[17] Two different books of criminal law were being followed. The one for Europeans was implemented in 1867 and the one for non-Europeans in 1873. By 1918, these two books would be combined into a single volume.[18]

The White Terror

This restrictive regulatory environment developed alongside the spread of anticommunist counterpropaganda. A document titled "The Methods of the White Terror to Suppress the Revolutionary Movement in Indonesia," written by Alinoeso—most likely an alias to refer to Alimin and Moeso—from Soerabaya on November 1, 1926, records a list of counterpropaganda that he calls "white terror."[19] One of the main threats came from Sarekat Hidjo (Union of the Green), which Alinoeso suspects was established by the Dutch government. "The government in cooperation with capitalist cliques and syndicates spent big sums of money for salaries, wages, and briberies. The union is composed of gangs of bandits and vagabonds to execute the dirty work." In Sumedang, with swords and batons, the "Greens" attacked a mass public meeting held by Sarekat Rajat (SR) and injured over fifty attendees. Leaders of the meeting and many communists were sent to jail, and forty of them were sentenced to three to six years of forced labor. During a public meeting held by the PKI to celebrate May Day in Chiamis in 1925, a fight broke out between the members of SR and the Greens (or "fascists"). At night, "a band of fascists numbering fifty to a hundred men went around, singing antirevolutionary songs, offending members and nonmembers of the party, who were against them." Young members of the PKI fought back and attacked this "annoying gang of fascists so severely" that they quickly disappeared.[20]

These anticommunist attacks did not occur only in these two West Javanese towns. In big cities and small towns all over Java and in other islands, communist members experienced similar violent threats. In Bandung, Soekaboemi, Semarang, Djepara, Rembang, Chepoe, Sidomulyo, Donomoelyo, Malong [Malang], and Baloeng in Java, as well as in Padang Panjang and Atjeh in Sumatra,[21] houses of party members were thrown with stones and dirt or were set on fire, and their properties were destroyed. Attacks, harassment, and molestations against communist members occurred daily. Several SR schools were burned down during the night.[22] Some members were badly beaten by the police.

> Women members of the party were arrested when visiting the meeting [in Japara], sent to the police station, locked up in one room to be humiliated. . . . A woman member of SR was arrested by 40 policemen and fascists and dishonored by them in a bestial way.[23]

The victims were not just male party members; women were targeted, and other races were affected in equal measure. In Rembang, "During the public meeting of SR, one Chinese and one member of the party (Malayan) were arrested by police, handcuffed and imprisoned for an unknown reason. They were beaten and mistreated badly."[24] The attacks were so brutal that members of the party who worked in places like Chepoe, which was the central of oil industry in Java belonging to the Royal Dutch Shell Company, and Baloeng had to seek refuge to other parts of Java.[25]

Repressions did not just come from the government; capitalist groups also spent large sums to suppress the *pergerakan merah*. A classified letter circulated among Suikersyndicaat (Sugar Syndicate) to Java Suiker Werkgevers Bond (Java Sugar Employers Association) "to prevent communist influence in sugar factories," an internal and classified regulation to surveil the workers. The letter advised administrators to gather *ledenlijst* (member lists) of the PKI and SR in each of their towns by consulting with the local government. They asked managers to cooperate with the Dactyloscopies Bureau, an office that kept photos and fingerprints of communist members to screen and identify workers—or potential workers—who were communist members. W. F. Wertheim argues that the foundation of the Dactyloscopical Bureau "enabled employers to fingerprint all laborers considered troublesome with the intention of excluding them from employment with any enterprise."[26] In other words, media technologies like photos and fingerprinting were used to manage, discipline, and punish workers. Sugar administrators were also asked to watch if their workers attended *blenggandering-blenggandering* (a Malayized term for *vergaderingen*) held by SR and the PKI. To surveil the influence of communism, they were suggested to hire a *spion* (spy) from among the workers. On Friday night as the weekend began, "it should be easy to watch" communist workers who left their houses to attend *blenggandering*. "Also watch if among the children of the workers, there are those who joined Sekolah Rajat [People's School] so as to predict the soul and political stance of the parents."[27] The involvement of sugar plantation workers in mobilizing the movements in rural plantations is important in light of the implementation of transport technologies to move both indentured labors and the flow of commodities, which enabled the Indies to become one of the world's leading sugar producers.[28] The need for the sugar companies to surveil their workers meant that their involvement in the movement had become a threat.

Nonparty members also fell victim to these daily conflicts, as did police officers. In Sidomulyo, during a protest meeting against the unlawful acts of the police and the "greens," as Alinoeso notes in this report, a brawl resulted in injuries among both communists and policemen. A European police officer was stabbed in the belly, leading to mass arrests. Forty communist members were put to trial and were sentenced to prison for stretches that spanned between two months and seven years. Likewise, in Padang Panjang, a native policeman was shot by a communist member while making arrests. Two similar incidents involving the killing of an officer occurred in the same town.[29]

Despite the many casualties involving Sarekat Hidjo, the coverage by mainstream Dutch newspapers was minimal. On November 23, 1925, *De Indische Courant* reported the role of Patih (vice regent) of Sumedang in Sarekat Hidjo to create a reaction against the popular movement led by the red Sarekat Islam (SI).[30] *De Sumatra Post* covered complaints that were made by two board members of SI Cianjur to the adviser of native affairs in Batavia, who reported reactionary experiences from the green organization.[31] The origin and nature of

Sarekat Hidjo remains a puzzle and its obscurity is equally reflected in existing literature that explain them only in passing.[32]

Conflicts became ordinary, as did arrests of communist members. In Belu island, 200 members of SR were arrested on the suspicion that they had organized an uprising.[33] Beginning in 1924, anticommunist conflicts were experienced on a daily basis, creating casualties not just of the ordinary party members and sympathizers but also nonparty members and police officers. Surveillance and control were also tightened in factories and plantations. Protecting public peace and order was a priority by the government, which was acting primarily to pacify the popular *pergerakan merah*. The Dutch state was taking actions through laws and regulations related to the communicative practices of the citizens to hinder and repress further development of the movement in the Indies.[34] It was in this period that the Dutch colonial state first took the arena of communication as a serious threat to its existence and invented and amended an elaborate set of legal measures and policing institutions. The legal measures by the state surrounding communicative practices became a part of the whole communicative environment that made up the history of the *pergerakan merah* specifically, and more generally of the broader Dutch colonization project.

The Rights of Association and Gathering

The rationales around the changes in the law shifted over time, as seen in the changing constitutional right to hold and attend OVs as a circuit of struggle. In the beginning, it was developed to recognize and protect human fundamental freedoms; however, soon as it was adopted in the colonial Indies, the law was implemented to regulate, control, and later repress political activities. The later changes became more obvious, especially in the period of the *pergerakan merah* in 1920s.

The constitutional right to OVs was protected under the Recht van Vereeniging en Vergadering (Dutch: the Right of Association and Assembly). Influenced by the spirit of the French Revolution, Recht van Vereeniging en Vergadering was first instituted in 1848 in the Netherlands under the protection of the constitution to recognize the rights of Dutch citizens to create political and cultural associations and to hold their assemblies.[35] However, while this law was developed to protect the Dutch citizens' rights for political activities, its adaptation in the colonies forbade these same political activities for the colonized.

Less than a decade later, in 1854, a regulation to prohibit—instead of recognize—the activities of associations and assembly in the Dutch East Indies was installed to prevent the potential of threats to "public order."[36] This prohibition of association and assembly was not lifted, even after the installment of the

Ethical Policy in the early 1900s, when the Dutch colonial state began to recognize the natives' rights regarding education, public health, and modern infrastructure. Still, native political organizations, inspired by the Eurasian and Chinese who had earlier founded their own political organizations, emerged, and developed on a large scale from the turn of the twentieth century onward, including SI and the PKI. The proliferation of these political organizations was supported partly by OVs, which served to organize and mobilize. First held on December 25, 1912, in Bandung by the leaders of the Indische Party Edward Douwes Dekker, Tjipto Mangoenkoesoemo, and Soewardi Soerjaningrat, *openbare vergaderingen* was later borrowed and popularized among the *kromo* by SI.[37]

This proliferation of native political organizations and OVs led to the change in the law. In the spirit of the Ethical Policy, the governor-general Idenburg recognized the right of citizens to associate and assemble in the Dutch East Indies in 1915 under the government rule of 111. However, while it was clear that the widespread presence of political organizations and public meetings made the government anxious about state security, it could not continue to ban them; in fact, despite the prohibition, the natives' political activities had expanded. The creation of new laws to regulate political freedoms was considered more suitable and beneficial for the government. Following this rule, a royal decree was released to ensure that the recognition of the right to assembly would not create a threat to public order. According to the state police guidebook, the regulation states that "the exercise of that right would be regulated and limited in the interest of public order." Together with this royal decree, the regulation of natives' political associations and assemblies was initiated on September 1, 1919.[38]

The rise of communism and the mushrooming of labor unions in 1919 sparked an increase in strikes and protests. However, two disastrous incidents that year became a turning point for Dutch ethical officials. The first was the Toli-Toli incident in Central Sulawesi, in which a Dutch controller was murdered a month after a propagandist from SI sparked mass refusal among local people to perform corvée obligations. A month later in West Java, Hadji Hassan and his family were shot dead by armed police at the direction of the assistant resident, an incident which uncovered the existence of a secret organization SI Bagian Kedoea (Second Branch) or Afdeeling B (B Branch) with "subversive" purposes.[39] Dutch ethical officials realized that the ethical thinking had opened a Pandora's box in which the campaign of freedom and transparency could be turned against the Dutch colonial state. Johan Paul van Limburg Stirum, who at first tolerated and respected the idea of Indonesian nationalism, took restrictive steps by adding in the royal decree of 1919 Article 8a to the Rights of Associations and Vergadering, allowing the governor-general to forbid meetings in certain areas.[40]

The right of assembly and association for the citizens in the Indies was accepted by the colonial state not to protect the natives' freedom of speech, but instead to control and restrict it. The motivation to control reflecting the government's anxiety around security is demonstrated in a passage from the royal decree in Stb. No. 1919, which states:

> The right of citizens to association and assembly shall be recognized and it was also determined that the exercise of that right would be regulated in the interest of public order in general and limited regulation.[41]

It was not made clear in the wording what "public order" meant, nor was there an explanation of what "general and limited regulation" entailed. But the article was followed with the following note:

Prohibited associations:
1. whose existence or purpose is kept secret;
2. which in accordance with the following article by the Supreme Court of the Dutch East Indies have been declared to be in conflict with public order.[42]

At the time, no association in the Indies was considered to be "prohibited." When the note was written, it pointed indirectly to a Chinese secret society in the Netherlands that had been banned earlier.[43]

Subsequent changes in the law around the right of association and assembly soon followed, starting with an amendment in 1923 that was written in response to the growing threat of the *pergerakan merah*. Two strikes held by communist-leaning organizations triggered government reaction. The first was a strike led by pawnshop workers in 1922 that led to the arrest and exile of Tan Malaka and A. Baars. The second was a strike by the railway workers in 1923 that led to the banishment of Semaoen. Both strikes were planned for several months in OVs in which Tan Malaka, Baars, and Semaoen were involved.

The 1923 amendment shows that while "public order" had been the main concern of the previous 1915 version of the law, regulations around "the exercise of that right" specifically addressed the activities in the movement that were considered a threat to the state. The amendment was created after the May strike led by the railway workers, which forced most local governments in Java to completely ban OVs. Although the ban was not lifted until October 1 of that year, an amendment of the regulation by the central government was already underway, which added three articles that led to limits in at least three areas of the conduct and participation in *vergaderingen*: nationalities, spaces, and age.

Section I (Association), article 2:

> Those other than Dutch East Indies nationals may not be a member of political associations and have to have attained the age of eighteen years. (As amended by Stbl. 1923 No. 452 in conjunction with 453.)[44]

The Dutch colonial state deemed that it was important to regulate the nationalities of the people who attended the public meetings for two reasons. First, the communist movement was founded by the Dutch socialists—Henk Sneevliet, A. Baars, and others—who brought and propagated socialist ideas to native political organizations and native organizers. This is apparent in the state's perception of the *pergerakan merah* as a movement that had been created from the outside and controlled and mobilized by foreign entities. By banning and regulating the nationalities of the people who could run and attend the meetings, the government believed it could prevent foreign influences from contaminating the natives. Second, the need to regulate nationalities was triggered by the active involvement of people of other nationalities.

Another new article, Article 5, mandated that all OVs had to receive a permit before the meeting was held, which helped the government to profile the meetings and the people involved, and to regulate the space in which they took place.

Section II (Meetings), article 5:

> Public meetings to debate are prohibited in the open air, unless prior authorization from the Head of Local Government has been obtained.
>
> Head of the regional State Administration may revoke similar permit or refusal of authorization by the Head of Local Government, bestow her on his part at the request of interested parties. (As amended by Stbl. 1923 No. 452 jo. 453.)[45]

By requiring the meeting organizers to obtain permits, local governments could have a say on which organizations that could hold a public meeting and track down which organizations held a meeting, where it took place, and the names of the speakers. The article was also followed by another restriction regarding space. Three kinds of *vergaderingen* were legally recognized: *openbare* (public in closed space), *openlucht* (public in open space, such as parks), and *besloten* (closed, and generally exclusive to membership card holders). The different meetings meant that diverse kinds of people could attend and listen to the rallies. *Openlucht* were seen by the state as the most dangerous because different people could come and access the meetings; therefore, the article specifically states that "public meetings to debate" could not take place "in the open air."

Further regulations, as reflected in Article 6 of the amendment, identified three new additional components: police watch, age of the attendees, and the right of the governor-general to limit the right to attend.

Section II (Meetings), article 6:

In all meetings in which the public is admitted, the officials or the police officers have full access.

Article 7a:

At the meetings, people who have the characteristics as under the age of eighteen years have no access.

Article 8a:

When any part of the Dutch East Indies is threatened by serious risk of public order disturbance, the Governor-General can provide that the exercise of the right of assembly be subject to the following limitations:

a. the provisions of Article 5 apply to all meetings open to the public as well as to meetings held in public places.
b. ... the Head of the District Administration is authorized to prohibit the holding of the meeting.[46]

Meetings after 1923 were characterized by the frequent presence of the police, and often more than two dozen officers would be in attendance (see chapter 3). The increase in the number of *vergaderingen* held meant an increasing demand on the government to hire more police officers. These officers were tasked with keeping watch on the movement, including taking notes of the names of the participants of the meetings, and police personnel hired from the rank of the natives multiplied, given their linguistic and cultural knowledge of the people they were required to watch.[47]

The rise of internal policing from within the native population was followed by controls on the age of attendees, who had to be eighteen years and over. This was particularly problematic because of the lack of written verifiable documents such as identification cards; there was no way to prove one's age except by the police judging one's physical appearance. Therefore, age could be determined solely from individual "characteristics." Because many Indonesian natives were short in stature, many people who were over eighteen were expelled from the meetings. In fact, a report done after the revolts of 1926 to 1927 shows that the average age of the arrested communists was thirty-one years, "an age which is reached 15 or 20 years after having passed schools."[48] This new regulation also marked a new cultural construct of, and therefore politicization of, the idea of "children." Where did the number eighteen, the age purportedly signifying the entrance to adulthood, come from? Many of the native Indonesians started to work at a very young age, mostly because they could not access basic education, and many people got married and had children before reaching eighteen. It was thus not a surprise that "children" joined the gatherings, and many of them participated because they already worked on plantations. By using this arbitrary age limit, the police could expel them from the meetings, limiting the possibility for the movement to include many young workers.

Another point was the right of the governor-general to authorize local governments to ban meetings, especially if they risked disturbing public order. The new amendment allowed the government to decide and control what was in fact the constitutional right of the people.

The next legal measure produced in response to the 1923 strike was specifically intended to restrict strikes—a type of regulation that had never previously existed. On August 7, 1923, Article 161 bis was released.[49] It stated:

> Any person, who has the intention of causing the disruption of public order or damage of life in terms of the domestic economy, knew or should have to feel that the damage to public order or damage to life in terms of economy in the country eventually will cause or promote more people to neglect it, or despite being given a lawful order to do the work that has been promised or according to his [working] obligations he should do it, will be punished with a maximum prison sentence of five years or a fine of as much as one thousand rupiah.[50]

The regulation restricts the calling and staging of strikes as a "disruption of public order or damage of life." In this case, life is defined as "domestic economy," which includes production (the case of plantation workers on strikes) and distribution (the case of transport workers on strikes). The article also permitted the arrest of individual workers who joined the strike based on their neglect of their "obligations" as workers. This regulation affected family members, as well as active strikers; even the act of collecting funds for families whose husbands were in jail due to their involvement in a strike was considered a disturbance to public order and economic life, and hence was prohibited.[51]

Despite the amendment of Article 111 and the release of the new Article 161 bis after the 1923 strike, the frequency and scope of the *vergaderingen* between 1924 and 1925 swelled considerably, forcing the government to take even more repressive measures around the holding of *vergaderingen*. At the beginning of 1925, strikes were held separately by hospital workers, printers, and dock workers in Semarang. Hundreds of leaders and movement members were arrested. In November 1925, 6,000 metal workers in Surabaya went on a strike in which seventy-five leaders of trade unions were arrested and imprisoned. Mardjohan and Darsono were arrested and banished from the Indies while Moeso and Alimin were able to escape and emigrate abroad. Ali Archam, the chairman of the Red Labor Secretariat, was banished to Digoel.[52] Soegono, the chairman of railway union, was arrested upon arriving from a trip abroad in Semarang and was imprisoned for a few months before being murdered by alleged "fascist clique" in jail.[53] The following are names of leaders who were banished either abroad or to isolated islands:

1. Sneevliet
2. Baars
3. Brandsteder
4. Kordenoord
5. Van Munster
6. Mevr. Sneevliet
7. Tan Malaka
8. Bergsma
9. Semaoen
10. Misbach
11. Zainoedin
12. Ali Archam
13. Mardjohan
14. Tan Tsiang Leng
15. H. Datoek Batoeah
16. Darsono
17. Soekindar
18. Soerat
19. Pranvirosardjono
20. Tedjomartojo
21. Sismodi

On November 20, 1925, the government released the "emergency ordinance" under Stb. No. 582, which allowed the governor-general "in the interest of public order to restrict political gatherings for one or more specific associations."[54] The ordinance made it clear that it was created to counter "specific" associations, including the PKI, SR, and other affiliated trade unions. As a response to the burgeoning communist OVs in 1924 to 1925, this ordinance was formed specifically to restrict *vergaderingen*, but more generally to curtail the *pergerakan merah*, isolating the communist affiliated parties and unions from the masses.

The creation of this ordinance opened the door to abuses of power. Sunario, a native lawyer, contested it, arguing that it could not be used to suppress the communist party because party members abided by the existing regulation in their political activities. A state of emergency could only be implemented in the "event of war and rebellion." Since the *pergerakan merah* did not create any public disturbance, the emergency ordinance could not be used and in fact was "illegal."[55]

Sunario's scrutiny of the state's abuse of power and restriction of speech cannot be underestimated, especially because the implementation of the regulation of *vergaderingen* was coupled with another regulation that targeted individual leaders. Article 156 read: "to anyone who deliberately declares hostile feelings, hates or humiliates groups within the population in the Dutch East Indies will be punished a maximum period of four years or a fine fl.300."[56] Many rank-and-file leaders were arrested on the basis of this article; in giving speeches in rallies and OVs, they were accused of spreading hatred against and humiliating the government, the capitalists, and/or (Dutch) colonialism.

The case of *woro* Ati—one of the prominent female leaders from Malang—that opens this chapter is an illustration of the use of Article 156. She was the first woman charged with speech offense. The incident took place in Donomoelyo, one of the towns that bore the brunt of interpropaganda violence. In this town, during one of the more serious fights between police officers and party members, one party member was killed, and two policemen were stabbed to death. Several other members were imprisoned and sentenced to years of forced labor.[57] Ms. Ati was charged with a violation of the state's law because, in her speech in one *vergadering*, she invited other women to not choose *prijaji* men as husbands, especially if it was only to gain a "comfortable" life. Instead, she advised women to join in the communist movement until they gained independence, with the implication that a "comfortable" life would come from their own freedom to choose who they would marry. What truly led *to woro* Ati's one-year jail sentence in her hometown Garut was that she humiliated the *prijaji* class—the government officials—and at the same time boldly stated that if we are "prevented and killed" we "will not give up," and thus allegedly incited violence. Not only was she considered to have intentionally declared hostile feelings toward government officials (the *prijaji*), she did so by raising "extreme" sentiments—to not give up even if one threatens to kill—which, according to the judge, could lead to "public order disturbance."[58] Ms. Ati was among the numerous leaders who were arrested without clear charges. In *Api* this was expressed as: "*tangkap dulu perkara belakangan*" (arrest first, charge comes later), in which people complained about how the government unfairly detained the people.[59] As the repression against the *pergerakan merah* heightened at the end of 1925, public gatherings ceased.

The Press Banning Ordinance

To think of the regulations of the press as a circuit of struggle also reveals that the state's battle against the *pergerakan merah* was manifested in the legal changes regarding the operation of the press. Mirjam Maters argues that government's regulation around the press in the Indies in this period could be divided into two steps. The first step took place between 1918 and 1920 with the expansion of the network that supervised newspapers. One manifestation of this was the development of the periodical *Inlandsche pers overzicht* (*Survey of Native Press*, IPO). IPO contains reports in four categories: Malay newspapers in Java, Malay newspapers outside of Java, Malay-Chinese newspapers, and Javanese newspapers with Malay rubrics. Categorization is also made based on themes, such as extremist newspapers, women's newspapers, and labor union's newspapers, and on issues, such as people's movement, youth movement, economy, and so on. IPO summarizes articles published in the most important newspapers. It was distributed to lower rank officials, members of Volksraad, the postal administration, attorney generals, and the heads of district courts. The periodical was produced using the method of stencil printing until sometime in 1921 when Balai Poestaka bought a printing press machine.[60] The second step began in 1920 when the government started to take explicit actions directed to communist propaganda. "The fear against communism highly determined government regulation that criminal measures were tightened not only in relation to the press but also to verbal propaganda," both affecting the rights of associations and gatherings and the rights to strikes.[61]

Like the rights of association and assembly, freedom of the press was recognized by the Dutch colonial government, especially during the Ethical Policy era. Enacted on April 8, 1856, the Drukpersreglement (Printing press regulations): 1) allowed censorship by asking any publication to send a copy to the government before it was released to the public; and 2) introduced criminal liability for defamation, insult, or slander against the king of the Netherlands, his family, as well as public officials.[62] In 1906, this ruling was amended and put an end to this censorship regulation. However, this should be seen less as an end to the restrictions on printing presses and more as a tepid relaxation of the law.

When the revolutionary press emerged and increased rapidly in numbers and social effects, the Dutch government deemed it necessary to create a stricter environment for the press. At the time, however, administrative intervention into press matters was not allowed legally, so the authorities released *haatzaai artikelen* (hatred-sowing articles) to protect public order.[63] These articles indicate that the promotion of hostility, hatred, and contempt against both the government of the Netherlands and the Netherlands Indies, as well as against horizontal groups within the society in the form of words, signs, and depictions, would be punished. Under the *haatzaai artikelen*, the government was able to arrest editors and authors on charges of promoting hostility and hatred against the government. Between 1918 and 1923, most of the editors of *Sinar Hindia* newspaper were arrested and/or sent in exile, leaving behind only a few others to run the paper. On April 16, 1926, because of the lack of editors—at that point only W. Kamsir and Nawawi remained—and the continuous harassment of its readers, *Sinar Hindia* was closed. Other revolutionary newspapers met a similar fate. By the end of 1925 and early 1926, most editors, including correspondents from small regions, had been arrested and either faced jail time or were sent into exile abroad. The continuous arrests and exiles of editorial members demonstrate the contentious relationship between these newspapers and the government.

The government also perceived the threat of communism as an "external influence" by foreign entities. Steps to curb communist propaganda coming from abroad included the restriction to import printed materials. The prohibition to import printed materials is specified in *Staatsblad van Nederlandsch-Indie* 1900, no. 317.[64] It states that

> With this the governor-general, to protect public order, can prohibit the import of printed materials from outside of the Dutch East Indies, with the exception of the Netherlands. . . . These materials have to be confiscated and destroyed. The confiscation and extermination have to be conducted by the police. Import-export and customs tax officials must hold the prohibited goods and report them to the head of the local government. Forbidden items in the postal and telegram services may not be delivered or sent to anyone else. With the word "*interdit*" (forbidden) the item must be returned to the office of origin, or detained. Every European who, after this announcement of the prohibition, imports, transfers, releases, or possesses the prohibited goods will be sentenced to one to six months in prison and a fine of twenty-five to 500 guilders, while Indonesians receive forced labor without being chained.[65]

TABLE 9. List of arrested and exiled *Sinar Hindia* editors, 1918–23

NAME	DATE	CAUSE
Darsono	Dec 16, 1918	Press offense
Semaoen	March 12, 1919	Press offense
Marco	1918	Press offense
Darsono	October 1920	Press offense
Mhd. Kasan	November 1920	Speech offense
Marco	1920	Press offense
Partoatmodjo	1920	Press offense
Sanjoto	Dec 1920	Speech offense
Marco	July 4, 1922	Press offense
Semaoen	Augustus 1923	Press offense, exile to the Netherlands

TABLE 10. List of arrested and exiled editors of various newspapers of the revolutionary press in the Indies, 1925–26

NEWSPAPERS	ARRESTED OR EXILED EDITORIAL MEMBERS
Proletar	Moeso
	Soedibio
Soeara Tambang	Nawawi Arief
	Idroes
Pemandangan Islam	Dr. Batoeah
	Djamaloedin Tamin
Njala	Dahlan
	Moh Sanoesi
	Alimin
	Gondhojoewono
	Tjempono
Djago-Djago!	Natar Zainoeddin
Doenia Achirat	H. S. S. parpatieh

This prohibition, however, did not clarify the nature of the printed materials that should be confiscated and destroyed. In the face of the rise of communism, van Limburg Stirum asked the director of justice to specify and augment this 1900 prohibition. Released on January 31, 1920, the prohibition detailed that printed goods that were deemed necessary to be confiscated and destroyed were those that "encouraged people to do punishable actions, or to resist or disrupt public authority in the Netherlands, in the Dutch East Indies, or elsewhere."[66] Despite this law, communists smuggled printed materials via shipping lines, tucking them inside sailors' baggage or among regular newspapers or periodicals. In Tanjung Priok port, 700 bags of letters were screened and scrutinized every week.[67] Postal officers used photos of book covers to easily intercept the forbidden books. However, the communists often disguised the publications, using innocent titles such as *Riak di danau* (*Ripple in the Lake*), *Mayat berdarah di kereta api* (*Bloody Corpse in the Train*), and *Kembang putih* (*White Flowers*), to cover publications and printed books that discussed Vladimir Lenin, Joseph Stalin, Leon Trotsky, the Kuomintang, Comintern, the Soviet Union, and the Bolshevik propaganda and revolution.[68]

In 1925 and 1926, communist influences could no longer be contained. Strikes and public meetings were held by printers in factories, by sailors and port workers in Semarang and Surabaya, and by railway workers and metal workers.[69] The government then released a couple laws, the antistrike law (161 bis), the antirevolutionary press law (151 bis and ter), as well as a law criminalizing a revolutionary propagandist through KB 20 March 1926 Article 153 bis and ter that began to be implemented on May 1.[70]

> In article 153 bis it is stated that those who intentionally in the form of words, writing or drawings advocate disruption of public order, overthrow or assault of power in the Netherlands or in the Netherlands Indies, or create an atmosphere for that, are punished with a maximum sentence of six years in prison or a maximum fine of three hundred guilders. In article 153 ter it is stated that those who spread the aforementioned matters are punished with a maximum imprisonment of five years and a maximum fine of 300 guilders.[71]

These articles were more powerful weapons than any previous criminal laws to punish political agitation; they were intended to punish acts of contempt and expressions of hatred, harassment, or hostility toward the government and certain groups or classes in the society.[72] Immediately after the installation of these two laws, three revolutionary newspapers, *Api*, *Njala*, and *Proletar*, ceased to publish. Many editors were arrested on account of seditious articles.[73]

After the communist revolts in 1926 to 1927, even though most of the revolutionary publications bearing the flag of communism had met their demise, the government sought to create even stricter controls over native presses. In 1931, just four years after the banning of communism, and for the first time in the history of the press in the Dutch East Indies, the government invented a new law specifically to control the press. This included 1) banning printing, publishing, and distribution of print works; 2) seizing presses and other materials used by printing companies; and 3) completely closing the buildings and the premises of printing presses. This was released as the Persbreidel Ordonnantie (Press Banning Ordinance) on September 7, which legally called for the "Protection of public order against undesirable periodicals and printed matter."[74] Instead of arresting individual editors or authors, this ordinance allowed the government to completely ban publications for a maximum of eight days in the interest of public order, as well as to confiscate and close buildings if deemed necessary:

Article 2 section (1):
> If the designation referred to previous article does not have the desired effect, the governor-general may, after hearing the Council of the Netherlands Indies, issue a ban on printing, publishing and distributing such print works, in the case of a newspaper for a maximum of eight days and in the case of other printed periodicals for a maximum of three times the period between the appearance of two consecutive issues.

Article 3 section (2):

> The head of administration takes immediate action to prevent the print-
> ing, publishing and distributing of the magazine for the duration of the
> ban, and can proceed with the seizure of printing presses and other
> materials used at the printing company and with the sealing of build-
> ings or premises where those objects are located. The head is allowed to
> this end, if necessary, to use the strong arm to provide the access to all
> enclosed places, including homes.[75]

During the *pergerakan merah*, the government did not have legal grounds to
stop the production of printing presses; instead, *haatzaai artikelen* were used to
arrest main editors and printers. At the time, seizing printing presses was also
considered impractical because only a small number of printing presses existed
in the Dutch East Indies. One printing machine could print several newspapers,
both European and Indonesian, so the confiscation of printing materials would
jeopardize more than one newspaper.[76] This lack of a legal basis to regulate the
press was considered by the government as a limitation, as new editors and
printers emerged to continue the production of the press despite the constant
arrests of editors. The release of the Press Banning Ordinance meant that the
government could engage in a broader form of persecution against the press.
The ordinance allowed the government to decide whether a publication was
deemed dangerous to the establishment of the state and threatened "public
order" and to entirely ban and close operations, as well as confiscate machinery
and technology.

The Wild School Ordinance

Unlike the regulation of association and assembly and of the press, the regulation
around the creation of private schools, which the government called "*wilde particulier schooltjes*" (wild private schools), was not enacted until 1923.[77] According
to Kenji Tsuchiya, reports on unsubsidized private schools began to appear in
colonial government documents in early 1920s especially in relation to Chinese
schools, Taman Siswa, and the SI school in Semarang.[78] "Wild Schools" referred
to educational institutions that were not held or subsidized by the government.
Indeed, regulation around "wild schools" did not exist until the dramatic growth
of SI schools (later, People's Schools) in the early 1920s led by Tan Malaka. On
March 28, 1923, Staatsblad van Nederlandsch-Indie 1923 no. 136 to regulate
the proliferation of "wild schools" was released.[79] This ordinance ruled that

nonsubsidized schools were required to register their schools to the government, that is to the head of regional administration. It also stated:

Article (1) section 5:
Teachers have to report and explain the lessons taught.

Article (2) section1:
Teachers have to allow government officials to sit in and observe the class.[80]

This ordinance required for the first time that teachers report both their identity and the content of their courses. It also made it possible for government officials to observe the conduct in classrooms. This requirement created a burden for teachers in the People's Schools, partly because they ran on donations and many teachers were volunteers. One teacher might teach math on one day and Dutch on another. It was difficult to report the kind of lessons taught because the only available forms of communication with the government were personal letters. Many teachers were arrested because of failure to report the content of all their courses. Another problem with letting government officials observe classes was that the law did not specifically state the maximum number of the officials allowed in any given observation; on many occasions, more than three officials sat in on a class. This ordinance effectively allowed the colonial government to dismantle the "wild schools" using legal measures that the natives had no power to challenge or resist.

After the communist revolts, the fear of communist uprising was so great that the effort to dismantle the "wild schools" was not enough, which led to a new ordinance with even stricter regulation of the schools. In 1932, the government released Staatsblad van Nederlandsch-Indie 1932 no. 494 ("Toezicht-ordonnantie particulier onderwijs") ("Supervision Ordinance for Private Education"), also known as Wilde Scholen Ordonnantie (the Wild Schools Ordinance), which took effect on October 1. It required that a permit was obtained before a private school was established and that all teachers working in unsubsidized schools to register themselves with the local authorities and undergo a quality check by an official state representative.[81] The inclusion of the term "public order" in the ordinance indicates that it was produced with the erstwhile communist People's Schools in mind.

> Teachers teaching education in unsubsidized schools had to receive a permit from the head of administration before they can teach. The permit has to be requested through a letter with three copies of a photo attached.

> This permit will only be released if the school is deemed not to dis-
> rupt public order and that the inspector of education states that the said
> teachers were not dangerous.[82]

Reading this regulation through the concept of circuit of struggle illuminates
how the colonial state perceived the educational institution as a serious potential
threat to public order. The ordinance's condition that new teachers obtain a permit from the head of the local government and hand in three copies of a photo
allowed teachers to be profiled and surveilled for security measures.

After the ordinance was released, it was followed with *Memorie van toelichting*
(Explanatory Statement), which stated that the rationale behind the release of the
ordinance should be that:

> The government should confine themselves to prevent and cut off the
> excess and the removal of serious abuses. It aims only to take away some
> disadvantages, which occurred in the private education provision, that
> are in practice . . . only to counter excesses.[83]

However, none of the twenty-one articles in this ordinance mention "excess,"
"abuses," or "disadvantages." Instead, all of the articles were directed at all teachers
and at "wild schools" in general. The ordinance was criticized from many fronts—
political parties and public figures—including the organization of Indonesian
students (Perhimpoenan Peladjar-Peladjar Indonesia) as opening the door to
abuses of power.[84] In fact, within the legal terms used in the ordinance, the words
"public order and peace" (article 4), "good name and reputation" (article 4), and
"public order" (article 13) were repeatedly used, and yet no protections against
the possibility of "excess" and "abuses" of state power against "wild schools" were
mentioned. The repeated mention of "public order and peace" was considered by
the Indonesian students to be dangerous given that it was an "elastic and vague
term," with potential to lead to legal confusion.[85] Indeed, even the booklet on
the right of association and assembly produced for police officers acknowledged
that the term "public order" was an "elastic and vague term."[86] While the police
booklet did not treat this term negatively, the Indonesian student association saw
it as a way for authorities to abuse their power and potentially lead to "the assault
of the *public justice*."[87] The protest against this ordinance was so intense that the
government repeal the law a few months later in February 1933.[88]

"Public Peace and Order" as a Masquerade

After the emergence of the *pergerakan merah* in the Dutch East Indies, which
ended in the revolts of 1926 to 1927, the areas of communication, notably

press freedom, the rights of assembly, and the right to provide education, were most affected by the government's repression. The press, public gatherings, and schools were the main communicative means through which the *pergerakan merah* mobilized and popularized anticolonial consciousness. Although the use of these peaceful communicative means was a novel step in the struggle against colonialism, it also forced the government to heighten control and surveillance over native conduct. These legal products were not born in a vacuum, but rather in a constant dialectical interplay between the colonial state and the anticolonial movement. As anticolonial struggles in the Indies shifted from warfare to communicative means of resistance, the government eschewed military action in favor of laws regulating public communication. In this case, the area of communication became one of the main circuits of struggle between the state and the movement. Hence, the emergence of the revolutionary communication and of the laws regulating them became the period's critical characteristics.

The state's repeated mention of "public peace and order" demonstrates how it perceived itself as continuously in danger, and how communication became one of the areas that posed a real threat to its sovereignty. In the aftermath of the revolts, the new legal measures allowed the government to watch and control every communicative action carried out by the natives—the authority to restrict teachers in nongovernment schools, to ban and confiscate press products and press machinery, and to prevent and ban political gatherings—in the name of protecting "public peace and order."

With the tight control of the government over the communicative practices of the native public, the Dutch colonial state sought to create a myth that public order had been maintained through the implementation of these new regulations. In reality, public protests against the government, including political movements, were disabled and disempowered through repressive control. In the name of "public peace and order," this evolution of the legal measures on communicative practices and technologies reflects the broader shift of the Dutch colonial state from an ethical state—a product of the Ethical Policy at the turn of the twentieth century—to a police state.

The history of capitalism and the state is closely intertwined with efforts to discipline and contain labor forces and unrest. The ordinance suppresses subversive communication practices. This also shows us that everyday communism is not mundane. In the case of the *pergerakan merah*, it is subversive and revolutionary, and it changed the course of colonial rule with its turn to repressive mechanism of discipline and control.

The emergence of law and policy in the specific areas of communication in this period created a new order in the Dutch East Indies and provides a new insight into how later postindependence Indonesian communication law and policy stems from legal products from this period.[89] This history reveals two points about the nature of communication law and policy in Indonesia. First, at the outset, communication law and policy were born out of the struggle between the state and the people.[90] Second, it was produced to limit and control freedom of expression and speech, rather than to facilitate and protect it.

At the conclusion of the communist revolts, in March 1927, surveillance activity was tightened and the government released *Politiek-Politioneele Overzichten* (*Political Police Review*) to the upper echelon of government officials. The report included the development of extremist movement, nationalist and Islamist movement, Chinese movement, and "native" labor movement.[91] In 1927, "extremist movement" or communist movement made up 17 percent of the report. In 1928, activities in this category went down to 9 percent, and by the 1930s it only comprised 3 percent.[92] While the government perhaps believed that they had successfully appeased the *pergerakan merah* in the Indies, Rudolf Mrázek shows that myriad spaces of enlightenment emerged in Digoel. The internees continued to read and write, created a library and schools for children, and held theater performances.[93] There were music and arts; the roarin' twenties had reached Digoel. In the meantime, outside of the camp, Indonesian communists continued to move clandestinely and to keep aflame the fire of red enlightenment, most notably, as the final chapter illustrates, by traveling the open seas.

THE OTHER LABOR OF CLANDESTINE SAILORS

The work of writing, no one could help. Only I should do it with my own hands . . . without a table and a chair [I am] writing above the sea.

—Timorman, "Korban-korban PARI" ("The Victims of PARI")

Revolutionary movements are not developed through grandeur. They are developed through mundane, tedious, and, at times, dangerous work. Djamaloedin Tamin knew this life-sacrificing challenge in both the viscera of his rebellious spirit and the toil of his enervate body. From his jail cell in Batavia in August 1933, Tamin wrote about his clandestine journey during his seven-year escape from the Indies under the alias Timorman in a memoir called *The Victims of PARI* (Party of the Republic of Indonesia).[1] Sober and undeterred, yet at times shaken, he chronicled the stories of Indonesian communist party members whose revolutionary venture abroad after fleeing the Indies following the revolts of 1926 to 1927 was either met with prison, death by suicide, or simply by leaving the movement entirely. Some of those who left were dissuaded by the looming dangers in the strictly policed Malayan peninsula; others were motivated to pursue economic activity for their own personal survival. As the person who provided a shelter for these exiles when they first arrived, Tamin learned the dangers inherent in forming a new cadre. He witnessed firsthand how some members had turned their back against the movement and worked for colonial intelligence. At the same time, he made sure that those who continued the struggle against Dutch colonialism were not forgotten. He recorded in detail the comrades who continued to disseminate communist literature in the Middle East on their way for pilgrimage to Mecca, as well as a small number of those who worked closely with Tamin as sailors. Tamin's own escapade as a seaman was the most captivating story among many; it opens a new perspective on the key role of sailors in mobilizing the *pergerakan merah*. Trying to make ends meet while fulfilling the revolutionary demands for mobility and communication, Tamin faced the dangers of the sea, persevered through perilous physical demands as a sailor, and tiptoed around undercover police, all while performing revolutionary work on the side. The story of these sailors is a story of courage and willpower as much as suffering and despair. It is a story of revolutionary labor in its human form.

This chapter examines the role of sailors as the prime movers of the *pergerakan merah* in international waters and argues that their involvement is best understood from the perspective of "the other labor." These sailors integrated communicative means and practices into their everyday duties in the sea, producing communication and mobility, strengthening the network of the revolutionary vanguard, ensuring that communication continued in times of heavy repression by the colonial states, and sending important documents, newspapers, letters, and pamphlets to or from the Indies. The everyday works and actions for movement building and mobilization in international waters reveal how revolutionary endeavors were produced as "the other labor"—an alternative kind of labor produced within working space and time that are repurposed and reappropriated for contentious politics. In this case, "the other labor" offers spaces of red enlightenment, expanding the war of pens and words and cultures of resistance within the vicissitude of the *pergerakan merah* going underground. Through "the other labor," sailors translated individual suffering into universal solidarity, turning the alienating and divisive conditions they experienced as sailors into a force of collective power. In short, sailors had a major part in creating intercontinental communication, connection, and exchanges.

While previous literature has rightly noted the international dimension of Indonesian communism, the work of many ordinary Indonesian sailors and port workers has not been fully appreciated. As recent publications show the Indonesian transport workers' involvement in revolutionary politics was not only central in radicalizing the communist anticolonial resistance in the Indies. It was also part of a broader international cycle of rebellions against global imperialism in the eighteenth century, which continued into the twentieth century.[2] This chapter discusses how the *pergerakan merah* did not completely disappear after it was banned following the 1926 to 1927 revolts. In fact, the movement continued clandestinely beyond the archipelago through sailors.

Sailing at Two Frontiers

The story of Indonesian sailors' involvement in the *pergerakan merah* started a few years before Tamin's clandestine activism as a communist sailor in Southeast Asian waters. The successful mass strike of Vereeniging voor Spoor-en Tramweg

Personeel (the Union for Railway and Tram Workers, VSTP) in early May of 1923 (discussed in chapter 3) led to the exile of the legendary leader Semaoen. Upon his subsequent arrival in the Netherlands, Semaoen was greeted by Dutch comrades who organized a public meeting—an *"openbare welkomst vergadering."* Amid a friendly and supportive crowd, Semaoen was carried on two men's shoulders as another man behind them carried a big poster that read:

OPENB. WELKOMST-
VERGADERING
In 't Concertgebouw, waar
SEMAOEN
De uit Indië verbannen arbeidersleider
Wordt begroet
Op
VRIJDAG 21 SEPTEMBER
SPREKERS:
SEMAOEN, KITSZ, BROMMERT, LANGKEMPER
AANVANG 8 UUR ENTREE 15 CENT
De vergadering wordt opgeluisterd door Zang en Muziek

A welcoming public meeting
In Concertgebouw, where
Semaoen
The union leader exiled from the Indies
Will be greeted on
Friday, September 21
Speakers:
Semaoen, Kitsz, Brommert, Langkemper
Begin at 8 entrance 15 cent
The meeting is graced by songs and music[3]

Bursting with music and songs, the September 21, 1923, meeting became both a celebration for Semaoen's achievements as a union leader and an opportunity to raise awareness about the struggle against Dutch imperialist actions abroad. The scene was so powerful that Semaoen was persuaded to stay in Holland and begin organizing its sailors, which at the time made up the largest population of working-class Indonesians in the country.

Within just a few months, in January 1924, the Sarekat Pegawai Laoet Indonesia (Union of Indonesian Seamen, SPLI) was founded. Based in Amsterdam, the SPLI was able to gather at least more than 1,200 members in its first year.[4] Letters from two Dutch steamship companies, Nederland and Rotterdamsche

Lloyd, along with some intelligence reports, confirmed that most Javanese sailors aboard their ships, including SS Insulinde, Patria, Tjeremai, Tambora, and Goentoer, were members of the SPLI. The number of Indonesian seamen and port workers who joined this communist union swelled in subsequent years. By 1925, its members included roughly 3,000 seamen and 2,000 dockers.[5] In February of the same year, the SPLI's Indonesian-based affiliate organization was created by unifying the existing seamen's and dockers' unions into one organization called Sarekat Pegawei Pelabuhan dan Lautan (Seamen's and Dockers' Union, SPPL).[6] Profintern, an international body established by the Communist International (Comintern) with the aim of coordinating communist activities within trade unions, noted in 1925 that the SPLI and SPPL combined consisted of 3,000 seamen and 9,000 dockers.[7] Soon after they were founded, SPPL and SPLI joined Profintern and maintained international contact through Profintern offices in Canton, Manila, and other ports.[8] In a time when most main leaders had been banished into exile, the involvement of these sailors in the SPLI facilitated communication between the Indies and Europe, and expanded the international network of solidarity against imperialism beyond Europe.

The Communist Party of Holland as well as the Comintern office in Russia welcomed the idea of Semaoen serving as the representative of Partai Komunis Indonesia (the Communist Party of Indonesia, PKI) in the Netherlands. Funded by Comintern, Semaoen founded the Holland bureau of the PKI in Amsterdam. The bureau was tasked with establishing regular communication with the movement in Java, to follow its development, and to supply it with news from Comintern. It was also in charge of taking care of regular information and reports about the Indies for Comintern and Profintern, which funded salary and the PKI's office in Amsterdam. These funds paid for the staff's wages and for the publication and organizing activities in SPLI. In the first few months after SPLI's foundation, the PKI branch in the Netherlands used an office space and publishing facility at the office of the Dutch Communist Party.[9]

The integration of Indonesian sailors and the SPLI with the global communist movements allowed for Indonesian communists to contribute and benefit from the existing communication infrastructures that belonged to Comintern. OMS (Otdel Mezhdunarodnykh Svyazey, the Department of International Communication) was established within Comintern in 1921 to organize liaisons between Moscow and the communist movements in each country to conduct these supervising activities.[10] Comintern established regional headquarters in Shanghai: the Far Eastern Bureau in 1926 and the Pan-Pacific Trade Union Secretariat in 1927. Both organizations supervised local communist movements in East and Southeast Asia, allocating funds, dispatching couriers and agents, and receiving local communists who sought training in Moscow. Shanghai was an ideal location, as it was also closely linked with Russia through the "Red Underground Communication Line," that was comprised of a clandestine network of Chinese communists forming a transportation route between Shanghai, Hong Kong, and the Central Soviet Region.[11] With its Shanghai-based office, the Comintern built and maintained "a complex East and Southeast Asian liaison network of agents and couriers who developed critical links with the Shanghai regional headquarters staff," as well as "liaison officers in the local communist movements who acted as the nodes of the network."[12] Along with Teo Yuen Foo and Nguyen Ai Quoc, Indonesian leader Tan Malaka was among the regional facilitators tasked to construct and maintain regional network. He would travel between the Dutch East Indies, British Malaya, the Philippines, and China to facilitate cooperation among revolutionary movements in Asia.[13]

The story of Indonesian sailors at the other frontier, on the other hand, was lonely. These sailors had moved away from Moscow and sought to propagate their own version of communism, which posed multiple dangers after the banning of communism in 1927. While the SPLI was a story of sailors who became communists—"sailors cum communists," the second story, narrated and experienced by Djamaloedin Tamin, was a story of communists becoming sailors—"communists cum sailors."

The *pergerakan merah* in the Indies was mobilized legally and publicly until the banning of *vergadering* was instituted on December 13, 1925, by the Dutch colonial government. This led to the cancelation of the PKI congress, which was to be held at the end of that month. After the ban, the PKI began to mobilize as an underground movement, and plotted a revolt, which was held in Bantam and Batavia in Java in November 1926, and in Silungkang, West Sumatra, in January 1927. After the revolts, thousands of communist members and leaders were arrested. Some, even those with no sailing experience, managed to flee abroad by taking jobs as sailors. Some came from wealthy families and would continue their trip to Mecca via India and Egypt. Most others, who came from cities like Jakarta, Bandung, Banten, Pekalongan, Jambi, Rengat, and West Sumatra, flocked to the port city of Singapore, which had become a base for several previous communist leaders who had been banished from the Indies. Singapore was a key hub that connected Europe to East Asia to Australia. Under the control of the British colony, nearly all ships from Europe, Australia, and the Indian Ocean to China and Japan would make a stop there, making it by far the most important port for colonial trades and beyond. As the communist sailors arrived in Singapore, they were quickly given shelter by Tamin.

Tamin was a former teacher at the Sumatra Thawalib school, a modernist school of Islam based in Padang Panjang, West Sumatra. At the height of the expansion of the communist movement in Java in 1923, two teachers, H. Datoek Batoeah and Tamin, introduced communism to the school and began to publish communist works, including the newspapers *Djago! Djago!* and *Pemandangan Islam*. Batoeah was later sent to exile to Alor due to his attempt to create a communist branch in Aceh, but Tamin would move on to become Malaka's "principal lieutenant."[14] It did not take long for communist teachings to attract the hearts of the students. The Thawalib school quickly became an important base for the *pergerakan merah* in Sumatra and, as in Java, comingled the beliefs of Islam and communism. However, the relationship between the communists and the Islamists in Java ended in a bitter split while, in West Sumatra and its Minangkabau diaspora, the relationship between the two continued to thrive.

The network of Thawalib school's graduates would later become a promising key to the continuation of the movement after the communist revolts. As the *pergerakan merah* was heavily suppressed in Java, the movement would continue clandestinely in the Malacca Strait, which connected Sumatra and Malaya. Thawalib disciples would travel to Mecca for the hajj or to further their Islamic education; others would land a job as an Islamic teacher in Southeast Asia, stretching the mainland from the neighboring Malayan peninsula to Siam, thanks to Thawalib school's exceptional reputation on its strong Islamic education. The movement of these students, disguised as a religious mission, served as a network of human messengers whose main task was smuggling letters, publications, and messages.

On June 2, 1927, Tamin joined the leader in exile Tan Malaka, who also originated from West Sumatra, along with another comrade, Subakat, in Bangkok to found Partai Republik Indonesia (PARI). Tan Malaka intended PARI to be "revolutionary and communist," which implied that it would replace the moribund PKI.[15] In reviving Indonesian communism, they agreed that PARI would serve as a critique against Moscow's hegemonic influence and leadership. Because of its devotion to Moscow, the PKI had, in their eyes, diverted its attention from the *kromo* people's interests. As early as 1924, Tamin had begun to declare that the PKI's main propaganda "expects everything from Moscow," adding that any criticism against Comintern would be dismissed as "stupid" by PKI fanatics. Tamin went on to liken the PKI to the white wing of Sarekat Islam (Islamic Union, SI) saying that they both treated "their holy land" fanatically; SI was devoted to Turkey and the PKI was devoted to Soviet Russia. The core problem, he continued, is "that both do not have confidence on themselves."[16] PARI would revive communism differently because it would return its leadership to the Indonesian workers and peasants.[17]

Tamin's view here mirrored his teacher's Tan Malaka's stance vis-à-vis Moscow. As one of the important founders and leaders of the *pergerakan merah*,

Malaka was adamant that communism should not be directed top down and instead, as he asserted to Lenin in the Fourth Congress of Comintern in 1922, it must express specific needs and local concerns.[18] Tan Malaka's diversion from Moscow found support among his wide followers, including Minangkabau nationalist leaders. Despite not identifying as communist, Mohammad Hatta's Partai Nasional Indonesia shared similar characteristics with PARI: It is nationalist, anticolonialist, and socialist. Tan Malaka's publications would be circulated among these new *pergerakan* leaders and followers, and mutual respect among them created a benefit for PARI, whose principal contribution in the Indies was the distribution of Tan Malaka's writings. The circulation of these writings especially benefited from personal and family ties, which crisscrossed between the parties. "The amorphous, conspiratorial, and underground character of the Pari makes it very difficult to track down its actual activities."[19]

PARI serves as a unique chapter in the history of the *pergerakan merah* in that its strength and survival in Sumatra in particular—in contrast with Java—made possible the longevity of the movement beyond the communist repression. This rebuts the current literature, which argues that the PKI was moribund on Java from 1927 through 1945; Sumatran activities were quite widespread during this time frame.[20] However, the break from Moscow meant that PARI leaders had to mobilize without the Comintern infrastructures. Within a heightened coordinated watch and persecution by the three colonial rules in the region, they had to move in secret. Within a month or two after the first wave of communist sailors arrived in Singapore, Tamin had to find them jobs. Even before the break of the revolt, the three colonies had tightened their restrictions against communism, letters were exchanged between British Malaya, Dutch East Indies, and French Indochina between April 1925 and December 1926, and an agreement was made to cooperate on alleviating the "red scare" via a concerted and systematic effort between the intelligence agencies of the three colonies.[21] Attaining and holding jobs would help the communists avoid suspicion by the British CID (Criminal Investigation Department, or plainclothes detectives).

The intimate relationship between colonial rules and anticolonial struggles transcends geographical boundaries. Although the Dutch did not have an authoritative presence in foreign port cities like Shanghai, they had to cooperate with Shanghai-based police forces, the International Settlement's Shanghai Municipal Police and the French Concession Police.[22] In fact, the French, British, and Dutch secret police forces "not only openly exchanged intelligence, but also frequently arrested and deported revolutionaries from each other's colonies."[23] The global maritime networks too became the sites of struggles. As sailors and port workers created a chain of messengers smuggling communication within the communist network, captains of the ships worked together with the Dutch colonial government and foreign allies in a "maritime surveillance project net-worked across Asia."[24] Kris Alexanderson observes that "[w]ith ships serving as active sites of colonial policing and surveillance, . . . [s]eamen exercised a danger-ous mobility in the eyes of Dutch authorities, fostering political networks target-ing the same colonial expansion oceanic connections they helped create."[25]

The initiatives between the colonial governments had probably begun earlier after the 1926 to 1927 revolts in Java and Sumatra. After that, the neighboring British and French colonial governments were concerned that similar disrup-tions would sweep through their areas as well, and the Malaya Intelligence of Singapore produced monthly reports that centered exclusively on communist activities.[26] A Malayan Bulletin of Political Intelligence released in June 1927 explained how nine Javanese communists were arrested in Mecca early in June of that year. "They were found to be carrying on Bolshevik propaganda among the pilgrims and had made arrangements to start a newspaper."[27] Two organizations were formed: the Society of Sheikhs of Indonesia, which was a pilgrim Brokers Association, and the Islamic Association of Indonesia. Both groups endeavored to conduct propaganda among Javanese pilgrims.[28] To avoid raising suspicions among Singaporean intelligence agents, Tamin made sure that the recent escap-ists integrated to their new societies by working.

While Tamin gave these exiles a shelter, he too had to work and raise money for their food and lodging. To manage the high rent, they would cramp together in a rental house, sometimes with other workers from Japan, Ceylon, and India.[29] Still, they had to constantly move around, change jobs, and separate from each other to avoid close surveillance.[30] Before Tan Malaka left Singapore at the end of 1926, he funded these expenses using the salary from his job as an editor in a local newspaper. Often, Tamin had to beg for money from his sailor friends, who received a good salary and could easily make donations. These sailors also helped Tamin find jobs for the escapees. As a transportation hub, Singapore had ample opportunities for sailors, and lodging houses and sailor agents were open days and nights to accept new applicants to work in small ships with routes within Asia and the Indies, or in larger ships with global routes. Some sailor duties were not terribly daunting, and the pay was relatively high. Yet the flexibility of the job, in which an activist could move from one ship to another, was the most attractive—and coincidentally beneficial—aspect. A sailor had to carry *monsterboekje* or *zakboekje* (Dutch, lit. pocketbook). Like a passport, this book, released by the shipping company, was equipped with a photo and fingerprint of the sailor and provided information on its holder, including their name, place and date of birth, and nationality.[31] With this document, a sailor could easily show the captain of the new ship that he already gained working experience in a different company.

For the communists who came from Java to sell their labor, Tamin sent them to sailor boarding houses. Within a month or two, they would get a job as a coolie, a sailor, a helmsman, or even a cook. The communists who came from West Sumatra, however, were not used to working as casual labors. Many of them were religious students studying in the Thawalib schools. When Tamin contacted his network of ulema (Islamic leaders) from across the Malaya in Johor, Negeri Sembilan, Selangor, Pahang, and Perak, they happily and gratefully took these Thawalib graduates and hired them to be Islamic teachers.[32] The Malay village of Kerajaan Negeri Sembilan was also where Tamin's West Sumatran confidant, Bagindo Tenek from Sumur Pariaman (alias "Daya Jusuf" and "Aliasih"), lived and kept all of Tamin's secret documents—letters, books, and brochures.[33] However, not all of the one hundred or so escapees found a job in the region. Those who had money would continue their pilgrimage to Mecca, spreading the tenets of communism among the hajj along the way. Tamin made sure they brought some communist leaflets (some authored by Tan Malaka), including "Massa actie in Indonesia" ("Mass action in Indonesia") "Naar de Republik Indonesia" ("To the Republic of Indonesia"), "De jonge geest/Semangat muda" ("The young mind"), and "Gutji maksiat kaum militer" ("The urn of the military's sins"). Some communist publications often had to take a detour to Mecca before finally arriving in the hands of the Indonesian people in the Indies. Other fugitives withdrew entirely from communism after Tamin found them a job. It is not easy to work in a movement, especially one in which a person is exposed to continuous danger of persecution, let alone when his personal livelihood is at stake. When asked why they joined, many gave the simple reason: "to fix their economic situation first."[34] Out of the one hundred people he helped find a job in Singapore, Tamin believed that ten of them could help revive the PKI. To these ten comrades, he said:

> the only way for us to return to Indonesia and to reconnect with our revolutionary people/proletariats that have lost their leaders, there is no other way than: becoming a sailor.[35]

By becoming a sailor, one could return and connect with people in the Indies, but one could also flee the Indies and built the movement from abroad, in mainland Southeast Asia in the case of Tamin, and in Mecca in the case of many Indonesian communists.

The stories of sailors at the two frontiers present stories of contrast. The SPLI, under the power and oversight of Comintern, was able to turn sailors into communists and managed to create international networks with other anticolonial struggles in other parts of the world. The works of sailors in producing communication and mobility were funded and facilitated by Comintern infrastructures. PARI, in contrast, implemented changes and transformation in communist radicalism after the ban of the PKI in 1927 and the successive break from Russia. With the absence of Comintern support, these regular jobs sustained the communists' immediate living needs as they continued their party building activities. The story of the SPLI is one of sailors successfully organizing and amassing thousands of members and sympathy, but Tamin's story is a story of survival, hardship, isolation, and jeopardy, and of continuous hiding and disguise. Both revolutionary acts were laborious works that took shape in the everyday life of the sailors. Amid challenges and hardship, they produced materials that sustained, challenged, and reinvented communism.

Contentious Communication in the Life of Sailors

The communists integrated the duties of party building into their everyday life as sailors who produced communicative means and practices as their main products. In time of heavy surveillance within the Indies and abroad, port workers and sailors worked as messengers in expanding the *pergerakan merah*. By the end of 1923, most of the main leaders of the Indonesian communist party—Tan Malaka, Semaoen, A. Baars, Henk Sneevliet, and Pieter Bergsma, among others—had been banished from the Indies. During their exile, they continued to communicate with the Comintern while living in different places; Sneevliet helped develop the communist party in China, Semaoen and Bergsma stayed in the Netherlands, and Tan Malaka moved from Berlin to Russia to Canton, and later to Manila and Singapore.[36] Eventually, the movement of leaders became restricted and highly surveilled, so other means to carry information were needed. Sailors smuggled letters, newspapers, and brochures to and from the leaders in the Indies and those in other regions. It was not uncommon for crew members to smuggle illegal things. James Rush notes that opium was regularly harbored on China-Java lines.[37] Newspapers that were smuggled to or from the Indies, which provided important knowledge on the development of the communist movement, included *Sinar Hindia*, *Pandoe merah* (*Red Guide*), and *Djankar* (*Anchor*), a monthly newspaper belonging to the SPPL.[38]

Organizing the union for the sailors was another important task, and the SPLI was founded to organize Indonesian sailors who had access to go back and forth to the Indies. Organizing these workers meant that connections and solidarity could be built not just with the movement in the Indies but also with other communist or anticolonial movements around the world. The union was not founded merely to demand a wage increase. In September 1924, Dutch political intelligence was able to gather some documents, including a letter written by an SPLI organizer in the Netherlands to his comrade Mandur Marsam, arguing that the Javanese sailors received the lowest wage compared to other "'boys' of different races."[39] The Dutch government used this letter to inquire the steamship companies Nederland and Rotterdamsche Lloyd about Javanese sailors' working conditions in their ships. Both companies argued that Javanese servants enjoyed relatively high-paying jobs with a monthly wage of f.20 to 25 in Nederland and f.18 to 25 in Rotterdamsche Lloyd. Their typical working hours were 12.5 to 13.5 hours per day, and they enjoyed ten hours of rest "just like the European civilian personnel." The job consisted of light activities and keeping watch. However, as stipulated in the booklet, the SPLI was not merely a union. "In this endeavor, the SPLI stands at the point of view of the class struggle. It firmly believes in the unity of the working class throughout the world and fosters fraternal feelings towards all workers who legally fight for the aforementioned purpose."[40] It aimed to overthrow capitalism and take over the means of production, including transportation.

This organizing work was facilitated through ongoing communication. Hence, to be an SPLI member—and to be a communist—meant to write reports, read news, mobilize chains of messengers, smuggle literature and letters, and to attend meetings. For the SPLI, books, newspapers, and public meetings (*openbare vergaderingen*) were used to educate the workers on communist teachings and knowledge, and to provide a discussion space. In the Netherlands, the SPLI's meetings were led by the president, and the secretary took note of all conversations that occurred in the meetings. A historical sketch written by Semaoen titled *Pak Matosin* about the leading involvement of a fictional character Matosin in SPLI Amsterdam and Rotterdam gave a description of how a public meeting was conducted. The story depicted a discussion and conversation between Indonesian sailors at Café Kries in Rotterdam.

> "To me, it is still not clear what it means by socialism," said Matosin.
> "Socialism is a socialist community," explained the speaker.
> "What is a community [*masyarakat*]?" asked Pak Jah, second foreman from Tambora ship.[41]

As the meeting went on, the participants discussed the definition of "community," "socialism," and "capitalist ownership," as well as the history of Russian revolutions and the future of transportation in communism. Included among the attendees, the sketch mentions the presence of Merto Ikram, the first foreman (*mandor*) from the *Tjimahi* ship; Pak Djen, the first foreman from the *Tambora* ship; Rais, the second foreman from the *Insulinde* ship; Mat Tahir, the first foreman from the *Insulinde* ship; Mat Ngarip the second foreman from the *Sendoro* ship; and Sapar (Waginah), the head of the pantry from the *Insulinde* ship.[42] This sketch demonstrates how learning new concepts and educating oneself on communist ideas and terms had also developed into part of becoming a communist.

The SPLI also published its own organ, *Pandoe merah*, which was edited by Semaoen in the Netherlands. *Pandoe merah* was among the most important documents smuggled into the Indies to ensure communication and transfer of knowledge between Russia and the Indies. Although funding was difficult to secure, the SPLI was able to operate mainly through membership fees. As stated in a letter dated April 1924, the Comintern agreed to send 3,000 gold rubles to support Semaoen's work in publishing *Pandoe merah* and in handling communication between Russia and the Indies through sailors, who were providing reports for Russia about the development in the Indies, organizing the movement in the Indies, and complementing it with news and information from Russia.

However, working for and with the Comintern was not without day-to-day challenges. The transfer of money from Russia to Semaoen's hands proved to be difficult. The Comintern sent the money through the Communist Party in Berlin and then from Berlin to the Communist Party in Holland. In Holland, it took months for Semaoen to ask for the money he had been promised from a Dutch comrade. Semaoen received the money only after writing several grievances alleging that the money had been stolen.[43]

Security posed another challenge. Within less than a year after its foundation, the movement of the SPLI was already being kept under close watch by the Politieke Inlichtingen Dienst, the Dutch intelligence group. In a September 10, 1924, letter to the executive Committee of Commintern Eastern Department, Semaoen complained that the union was in dire need of an office space. They had been using the sleeping houses of the seamen to hold meetings, however, "the sharper control in the *sleepinghaus* of our seamen-comrades (belonging to the Steamer-Compagni) makes it impossible to meet our comrades illegally again there."[44] While the movement improved its organization and mobilization abroad, the Dutch intelligence heightened and expanded its watch and control both in the Indies and in the Netherlands.

By this point, the government was becoming increasingly anxious about the association that the SPLI was making with other international organizations. The SPLI's connection to the International Transport Workers Federation and the International Metalworkers, according to a Dutch intelligence report, meant that the Russian government had "information and courier service to control the production and transport of weapons."[45] It is unclear if this allegation was true, but the idea to organize maritime workers had been an interest of international communism for some time. The organizing of international maritime workers ensured coordinated actions among them and resulted in a better network of communication. The important role of sailors in creating global solidarity is nicely summed up in a poster made by the Communist Party of Holland that reads: "*Indonesië los van Holland nu*" ("Free Indonesia from Holland now"), followed with Karl Marx's statement "*Geen volk is vrij da teen ander volk onderdrukt*" ("No nation is free if it oppresses another nation") and a line urging to "*Kiest communisten*" ("Choose communism"). The poster depicts four sailors wearing their uniforms and hats, a mugshot of Marx, and a ship showing sailors' important role in the global solidarity against colonialism and capitalism.[46] The choice of sailors as a symbol for global solidarity reveals how the international network through conferences was also built by the Indonesian communists.[47] In his fictional novel, Semaoen recalled that Pak Matosin and Samsi were SPLI delegates to the congress of the union for world's revolutionary sea transport workers held in Hamburg in 1925.[48] A year earlier, in June 1924, the Javanese sailors joined in the Conference of Transport Workers in Canton, which had been instigated by the Red International of Trade Unions (Profintern). During the conference, the attendees praised China as the only place where the people of the oppressed could freely meet to discuss their oppression, and they thanked Dr. Sun Yat-sen for his effort to free the East from imperialism.[49] Attended by delegates from Java (Alimin and Budisutjitro), North and South China, and the Philippines, the conference published a manifesto addressed to the workers of Europe and America and the oppressed Eastern countries explaining how the workers of the East were awakened to organize workers' and farmers' unions, to form popular parties, and to call transport workers in the area of the South China Seas to join these organizations. This meeting was seen as the link to unite the revolutionary national liberation movements in the East with the class struggle of the proletariats in the West. In their words, "We call on all organizations of transport workers in colonial and semi-colonial countries to unite themselves and join the revolutionary transport workers of the world."[50] The conference lasted six days and the attendees agreed to establish the Bureau of the Red Eastern Labour Union in Canton, uniting all transport workers of all the Eastern countries and to create secretariats in China, Indonesia, the Philippines, Japan, and India. According to a Dutch intelligence report, following this meeting, the SPLI had planned to join a mass political strike simultaneously in Dutch and Indian ports.[51]

The work of organizing could also mean completely leaving one's job as a sailor. In 1925, three Minahasan PKI members who worked as cargo loading clerks for Koninklijke Paketvaart-Maatschappij (KPM), Djohannes Waworoentoe (alias Minahasa), Daniel Kamoe (alias Celebes), and Clemens Wentoek (alias Passi), left the ship in Rotterdam and joined two other sailors, Mohamad Saleh (alias Mulia) from Tapanuli and Oesman (alias Bangka) from Padang. They took the Russian steamer *Wasla Warasley* en route to Russia. Together with a sixth person, Soemantry (alias Dingli), a law student at Leiden University from Java, who took the train to Russia, they joined the University of the Toilers of the East (KUTV), a Moscow-based training college for communist cadres in the colonial world.[52] By joining the KUTV, they aimed to expand the movement in the Indies.[53] However, this effort too had to be recognized as a form of labor in which they had to overcome difficulties related to climate change, cultural difference, and university ethics.

FIGURE 9A. Taken from a travel card to Moscow is a photo of Daniel Kamoe.
Source: Executive Committee of the Comintern (ECCI) (1919–1943), box: F. 495, op. 214 (Personal Affairs [Indonesia]), case: 74, 1. 1, Russian State Archive of Socio-Political History (RGASPI), Moscow.

FIGURE 9B. Taken from a travel card to Moscow is a photo of Mohamad Saleh.
Source: ECCI (1919–1943), box: F. 495, op. 214 (Personal Affairs [Indonesia]), case: 75, 1. 1, RGASPI, Moscow.

A testimony acquired from an arrested informant claiming to be Kamoe mentions that roughly 1,000 students were enrolled at the KUTV.[54] The majority of the students were Russians, but a few came from China, British India, and Europe. Each class had twenty-five students, and they began at 7:30 a.m. with gymnastics and continued with a three-hour session on the theory of communism. In the first two months, the Indonesian students were given lessons by Semaoen on Indonesia and its political situation. After two months, most lessons in their first year focused on acquiring Russian language even though, as the informant describes, "[a]t the end of the first year, we still only understood very little of this language."[55] In the three years that followed, these students were given exclusive lectures on communism and the difference between communism and socialism, as well as gymnastics and military drills and exercises. Most student life occurred in a building that consisted of dorm rooms and classrooms. Each dorm room housed six to ten students of different nationalities. They received a fixed amount of 1.50 rubles for refreshments.

For these former sailors, adjusting to university expectations was challenging. Most of them came from peasant families; in the Indies, their parents cultivated land, and their siblings either helped the parents or stayed at home. They too worked as peasants before becoming sailors. However, despite little preparation in Russian language and culture and no experience with university life, most of them, except for Waworoentoe and Oesman, managed to perform well in their classes. Their education included discipline and active participation in party group (*kruzhok*) meetings, adopting the theory of Leninism and other party related disciplines and practicing this knowledge in university life, and conducting organizational work.[56] Soemantry—perhaps the most successful among them—had already gained law education and training at Leiden University, managed to write a thesis in French titled "Le mouvement anarcho-syndicaliste en Indonésie" ("The Anarcho-Syndicalist Movement in Indonesia").[57] Waworoentoe, however, missed classes without serious reasons and was not attentive in class discussions. Being "the oldest of all Malay Members," he probably felt the pressure to be a role model for the younger ones, but health conditions likely impaired him. In the two and a half years he was in Moscow, Waworoentoe suffered from chronic lymphadenitis, face erysipelas, and fibrous tuberculosis. Oesman, despite showing interest, had a weak understanding of political issues, barely comprehended lectures and course materials, and showed little progress. In social life, these sailors' relationships with their colleagues were varied. Mulia was praised as a good administrative worker in the housing committee while the two Minahasan Kamoe and Waworoentoe were reported to have "too much zeal for flirting with women."[58]

After these first Indonesian sailors were sent to Moscow to study at the communist school KUTV, more Indonesians followed suit. In a letter written sometime in August 1935 addressed to the communist leader Roestam Effendie, four Indonesians asked for Effendie's help to study in Moscow for longer than the six months that they had completed. It took them 1.5 years to reach Moscow in the hopes of reviving the PKI that had been destroyed in 1926 to 1927. They wanted to deepen their knowledge about the "theories needed for our movement" because a school employee had informed them that they should get ready to be sent back to Indonesia. Since Effendie worked in the Communist Party in the Netherlands, the students believed that he could help change this decision.[59] It is not clear if they were able to stay, but this letter indicates that there were more Indonesians who came to study at the KUTV after the first cohort of sailors.

The Other Labor: Communicating to Resist

The everyday works and actions of Indonesian sailors for party building in international waters reveal how revolutionary endeavors were produced as "the other labor," an alternative kind of labor produced within working space and time that are repurposed and re-appropriated for contentious politics. In time of the repression of the *pergerakan merah* in the Indies, to survive as a communist and to continue the work of party building meant to work as a sailor. Being a sailor became a condition of possibility for the mobility of people and messages. Besides acting as propagandists, sailors also became human chains, sending letters and smuggling communist publications, pamphlets, and newspapers. During a police investigation, the itinerant communist Tamin mentioned dozens of names (and aliases) that were parts of the chains of messengers who sent messages between and within mainland Southeast Asia and the Indies. Some were assigned to bring letters to the post office, some served as couriers to transfer letters directly to the leaders, and others sent news verbally. By going to the post office, smuggling communist literature, and handing letters to the addressees, these messengers repurposed their normal time and space to create an alternative space of resistance.

In charge of sharing information from and to Indonesia with correspondents in Medan, Pekan Baru, Palembang, Jakarta, Semarang, and Surabaya, Tamin knew that writing reports, gathering information on the movement from newspapers, and sending letters were important, albeit mundane, works that he had to do for the movement. His work as a sailor was merely a means to an end, that is to produce his political work. In his memoir, he wrote that on December 10, 1928, as he was working in the British-North Borneo ship,

> Two weeks we sailed, a week in Singapore, that's how it went for about a year. The one week we had in Singapore, we used the night time for reading letters, newspapers and meeting with colleagues, while from dusk to dawn we had to work really hard [in the ship]. If the ship was already sailing in the big sea, then we can sleep a bit. Even at night we had to work! Sometimes in those sailings, we had to work in the middle of the night because the ship made frequent stops and unload cargos. For two to three hours at night, one of us went all the way up to the pool or up on the ship's topmost deck to do night watch, to see if there are other ships, islands, etc. This responsibility to do the night watch is really hard. If the ship hit another ship or went ashore, then the people doing the night watch will be in trouble. Even during these times, we multitasked and wrote letters, reports, and read the newspapers.[60]

This excerpt explains the dual functions of labor enacted in the work of sailors. On the one hand, they fulfilled their job to do the night watch, to supervise the unloading of cargos, and to ensure the safety of the ships. On the other hand, despite difficult demands and work conditions, sailors also produced a different kind of labor. It is an act of labor, "the other labor," that sought to liberate the alienation of sailors, and other workers, from the exploitative working conditions under capitalism.

This idea of "the other labor" is best understood as an opposite of Marx's "estranged labor." In the context of colonial capitalist mode of production, labor was organized around colonial relations to work out the lands and labor in the colonies to generate and accumulate profit for the metropole. In this system, a worker appropriates nature by selling his labor. However, the object that he produces, which is a realized form of his labor, alienates him as something independent from him. This is because the fruit of the worker's labor does not belong to him but to the owner of the factory. In the case of the sailors, mobility produced by the seamen did not belong to them as an expression of life necessity in the truest sense but to the owner of the shipping companies.[61] This act of appropriation of nature within this system is called "estrangement" because as the worker appropriates the external sensuous world, he deprives himself of the very means of life and the means of physical subsistence mentioned above. In Marx's words,

> Thus in this double respect the worker becomes a slave of his object, first, in that he receives an *object of labour*, i.e., in that he receives *work*; and secondly, in that he receives *means of subsistence*. Therefore, it

enables him to exist, first, as a *workers*; and, second, as a *physical subject*. The extremity of this bondage is that it is only as a *worker* that he continues to maintain himself as a *physical subject*, and that it is only as a *physical subject* that he is a *worker*."[62]

This estrangement of labor is therefore constituted in three ways. First, the worker is alienated from both his labor, which itself is seen as a product to sell, and from the product of his labor. Second, "If then the product of labour is alienation, production itself must be active alienation, the alienation of activity, the activity of alienation."[63] In this case, the worker is alienated from the act of producing itself. In the act of a supposed expression of life, the worker is instead involved in an act of "*self-estrangement*."[64] Third, "In estranging from man (1) nature, and (2) himself, his own active functions, his life-activity, estranged labor estranges the *species* from man. It turns for him the *life of the species* into a means of individual life."[65] In short, the worker is alienated from his universal nature as a species being; he is denied of his life-activity. His labor, its product and his expression as a species being do not belong to him, but instead to his master, the capitalist.[66]

To conceptualize sailors' resistance against this condition of estrangement, labor should be viewed in terms of its duality, the estranged labor versus the "other" labor. The other labor is produced by workers through repurposing their otherwise estranging labor time and space. In other words, resistance is work that is repurposed as it is produced within labor time and space. While estranged labor (re)produced the colonial and capitalist system, the other labor produced the resistance against, and to transform, it. The other labor, that is the work of organizing resistance, that these sailors produced alongside their estranged labor generates different results. First, in producing the mobility for their master, sailors create "contentious mobility." Second, in the act of production, in the very act of conquering nature, sailors alienate themselves. Sailing is not an expression of his creativity but rather a means for his subsistence. Nevertheless, as they were negated, they expressed their creativity in the act of organizing. In negating the negation of estranged labor to their nature as species being, sailors mobilized a movement that appealed to universal emancipation.[67] In resisting the divisive system of capitalism that separated people based on race, nations, and class, sailors created solidarity.[68]

Using the same time and space in which he had to fulfill his night watch duty, Tamin wrote letters and reports and read newspapers to continue spreading information for the resistance movement. Constantly moving from one place to another, Tamin had to create opportunities to write, even if it meant writing above the sea without a desk and a chair.[69] E. P. Thompson explains how time measurement becomes a means of labor exploitation; in this case the same time measurement was repurposed by Tamin to be a means of resistance against such exploitation.[70] Hence, it makes a historical record regarding labor "not a simple one of neutral and inevitable . . . but is also one of exploitation and of resistance to exploitation."[71]

Thinking of revolutionary work as "the other labor" allows us to see writing as a microcosm of a whole social relation. The act of writing in the eye of the revolutionaries became an act of resistance that is often praised as a revolutionary character in itself. Tamin wrote at length about this nature of writing when talking about Subakat, the cofounder of PARI and the *Api* newspaper editor from 1924 to 1925 before he fled the Indies upon facing eight counts of press crimes.

> When [Subakat] felt stuck in the teeming crowd of a *vergadering*, he rushed to his desk and unsheathed his sharp pen filled of truth. Rebutting with the edge of his pen, rarely a pen expert [*ahli pena*] could wrestle with him. Rarely can people defeat him.[72]

A *kromo* and a chain smoker, Subakat would defend truth in writing with wit and humor.[73] In Singapore, Subakat worked as a tailor making Malay caps (*songkok*) made of velvet while propagating PARI and translating Tan Malaka's brochures, including "Massa actie." He was a conscientious translator who made sure he interpreted a sentence clearly and precisely, and this work was both tedious and time consuming. Like Tan Malaka, Subakat had to stay mobile to avoid intelligence. In one trip to Canton, he managed to see leaders from Moscow and reported the situation in Indonesia. "His newly gained English skills during the time as a fugitive was put into practice."[74]

As writing was adopted to be a revolutionary and emancipating labor, its products—letters, brochures, and books—became congealed forms of resistance. The fact that they were products of social relations and conflicts meant that they could become symbols of victory or pose a threat to the enemies. The colonial government's aggressive ban on Tan Malaka's writing and the intelligence office's aggressive tracking of the revolutionaries' writing artifacts are consequences of books as congealed products of social conflicts.

The development of colonial intelligence was closely related to that of the revolutionaries. Both revolved around writing and control over it, in a form of censorship, surveillance, search, and confiscation. Subakat was arrested by the British intelligence agency after they found his letters, many of which were addressed to the Communist Parties in the United States and France.[75] A quick learner and a witty, stubborn, yet humorous propagandist, he was sent to Batavia and died by suicide in jail. As Subakat was caught by following his letters, so was Tamin. When Tamin was finally caught by the British intelligence office in Singapore on September 13, 1932, he was presented with all the letters they had acquired in the post offices. One intelligence officer told him

> we know you always stayed in your room to read and write, and you usually wrote until the wee hours of the night. In the past two months, we were able to tightly control the movement of all letters "to" and "from" you. Sometimes we let the letters get to your hand so we knew exactly the nature of your work. We wanted to know how you would try to escape from us [if you were to be caught]. We knew you planned to go to Java on September 16, 1932, so we quickly arrested you.[76]

Both the act of writing and the products of writing became the biggest clue for the intelligence who had attempted to find Tamin for years since the revolt break in Java in 1926. The head of the British CID in the straits and the Malay Peninsula, A. H. Dickinson, told him several times that it was difficult to catch him. "It has been such a long time since we tried to find you, now we finally met. I reckon you had many friends who protect you."[77] Tamin had previously skillfully escaped numerous arrests by hiding, being in disguise, and as a sailor, traveling to Siam, British Borneo, Bangkok, Mindanao, Brunei Darussalam, and many other ports. If he saw an intelligence officer as his ship dropped its anchor in the Singapore's port, he would directly move to another ship to sail back into the sea. By taking control of the post offices, the intelligence was able to tightly manage the mobility of his revolutionary writings and map his roundabout.

As a microcosm of social conflicts, writing as the other labor carries a story of threat, failure, fear, and despair. In the jail cell, Tamin, much to his chagrin, recalled the surreptitious content of his letters posing dangers not just to himself but to others, many of whom were innocent, whose names he had mentioned. Eleven others were arrested related to his case, some of whom were fellow sailors who had nothing to do with the movement. On this he recounts, "I feel deeply sorry that I could not even swallow water. I feel remorseful of my writing and my action since many of those writing posed dangers and they [the efforts] were gone in vain."[78] For a few days, Tamin was so depressed that he would not eat or drink. On September 25, 1932, Tamin was charged by the British government for "holding a communist secret network to revolt against the British government and to propagate a similar spirit in other areas in the Straits and Malaya, Burma, Siam, Indochina, the Philippines, and Indonesia and to make Singapore a hub for such a network." Tamin pleaded guilty and admitted that he was the leader of PARI, but repudiated the scope of the movement, saying that his interest was merely "to propagate PARI in Indonesia among Indonesian proletariat."[79] Upon his arrest, Tamin, along with the evidence, was handed over to the Dutch intelligence in the Indies.

From Individual Suffering to Universal Solidarity

Through the other labor, sailors translated individual suffering into universal solidarity, turning alienating experiences into a force of collective power. Unlike estranged labor, in which sailors merely worked for a wage, the other labor rewarded these sailors with an affective infrastructure.[80] The idea of communism as a symbolic collective force and site inspired political confidence and practical optimism. It allowed sailors to think beyond their individual interest and to struggle for something greater than their immediate needs. The other labor, motivated by communism as the affective infrastructure, became the conduit that turned suffering into hope, fear into faith, alienation into collectivity, and oppression into freedom. It is work that turn a sailor from a mere worker into a human.

The suffering that sailors and port workers experienced was rooted in the condition of their labor. As one of the main ports in the region, Tanjung Priok had high traffic and flow of commodity from both the outer islands as well as Europe. This created demands for dockside coolies. Most of them were seasonal laborers recruited in remote villages in Java by foreman (*mandurs*) or helmsmen (*jurumudis*).[81] In Joseph Norbert Frans Marie à Campo's analysis,

> They did not want permanent housing, but they certainly did need temporary accommodation. In the old ports they stayed in slums in the polluted, unhealthy lower city. The new ports were kilometres away from the city, and traveling to and from the city was expensive and time-consuming. Anyway, the common practice of loading and unloading at night prevented them from staying away for long.[82]

As a result, slum neighborhoods emerged near the ports. Without drainage or piped water, they were susceptible to contagious diseases like plague, cholera, and malaria. These seasonal laborers would cram in with ten or even more than twenty people in these slums. Though some others might decide to spend the night among the cargos, these poor conditions made working in the ports "literally killing."[83]

The alienating condition of port cities could further be felt in the words of the Dutch romantic poet Jan Slauerhoof.[84]

Priok

Kapal hitam perlahan-lahan disorongkan ke dalam
Galangan yang dalam di antara los-los Panjang,

Pelabuhan-pelabuhan yang kaku memiliki segala yang beku
Dari kubur-kubur kota besar yang bertambah padat selalu.[85]

[The black ship is slowly pushed in
The deep yard between the long docks,
Rigid ports have everything frozen
From the tombs of big cities that are getting denser always.]

Here he describes ports as rigid and everything to do with them as "frozen"—stagnant in their poor uncomfortable situation, seemingly unable to be fixed. He called ports as the graveyard of big cities that were always becoming dense.

Due to this "coolie problem," two coolie kampongs (villages) were constructed in Tanjung Priok in 1918 and in Belawan Deli in 1928. Equipped with a central kitchen, constant medical inspection, a mosque and a cinema, their living conditions improved and became healthier than the previous coolie slums. However, the privilege of living in these kampongs required the coolies to be on call. In a way, the coolie kampongs benefited the shipping companies more than the coolies since it ensured around-the-clock labor.[86] The lives of these dockside coolies were lives of precariat labors, constantly lacking stability and security. For sailors, the housing situation might be slightly better especially when they arrived in ports abroad. In Singapore and Holland, they would be able to live in lodging houses. But they too were constantly faced with the danger in the sea.

In a way, both sailors and port workers lived in an alienating and unstable condition with working hours in which night became days and days became night. Moving from place to place and living far away from family, they had to redefine "home." This is vastly different from farming, where labors settle, get married, raise families, and create a routine. The sailor's working conditions are inherently dangerous. In the sea, dangers were fleeting moments that came unannounced yet could face them with physical threat. Routines were continuously disrupted. Perhaps they found routines in these disruptions; without family, they had to seek diverse sources of support in fellow sailors and port workers. Living in a small confinement with sailors from different races means they had to go beyond their individual source of comfort—race, nationality, language—and transcend those differences. Communism gave the symbolic tie and shared language of struggle between them. To be a communist gives these sailors an idea of "home."

Clandestine communication and international communist solidarity must be understood within this context of the precariousness of sailors and port workers' labor, as well as its alienating, unstable, and dangerous conditions. In that context, the sailor activists were able to repurpose time, space, energy, and money to perform a different kind of labor: organizing and building solidarity. In that effort of organizing, they moved resistance from personal to social and in that process changed their views about the world. Yet often, as Linebaugh and Rediker observe, "[t]he struggle against confinement led to a consciousness of freedom, which was in turn transformed into the revolutionary discussion of human rights" among sailors.[87] In the case of the Indonesian sailors and port workers, the immediate suffering experienced in the colony and at the workplace was discussed in the language of red enlightenment, in terms of "freedom," "liberty," and "emancipation" from colonization. This is captured so vividly in Semaoen's story of Matosin, who, upon fleeing his ship to stay in the Netherlands, fell in love with a Dutch woman, which led to an internal conflict. On the one hand he felt connected with the woman's family, who came from the same working-class background. On the other hand, she also came from the race of the people who colonized Indonesia.[88] This conflict was sorted through during their marriage, and Matosin changed his view about race and class, and more importantly humanity.

> [H]is strong will to guarantee his family life is strengthened with a firm will to help the destitute among the Dutch people as much as he can. A spirit grew in Matosin's heart to dedicate himself for the poor. In that spirit he does not see anymore the difference between nations; what he sees are just humans.[89]

As Semaoen, through the personification of Matosin, wanted to change the society, he also changed his own ideas about who the Dutch were, what race and class meant, what marriage could be, and what kind of relationship he should have with the Dutch people. The ability to change his view about race is central in redefining his idea of inequality and to take human causes as universal causes. In discussing the five sailors, Semaoen described them as "without thinking about the consequences for theirself [sic] and only wanting to fight with knowledge of the communism and to die for our movement, want to go to Moskow and to fight from their steamers."[90] Here the sailors were described putting their self-interests aside for the purpose of a larger goal. Similarly, Tamin wrote that a revolutionary was a person "who believes fully in sacrificing his life to continue the struggle for freedom."[91] While the story of sailors lacks romance and family relationships, it is still a story of devotion and commitment to revolution and sacrifice for the freedom of one's nation and, more universally, for humanity.

Revolution as the Work of Many

Recognizing the tedious day-to-day job of party building as labor allows us to unearth the role of sailors in the day-to-day process of developing a movement, which is often hidden when we focus on the narrative of a legend. In the popular imagination, for example, Tan Malaka inspired a fictional character of Scarlet Pimpernel, depicting him as fluent in several languages with "the uncanny ability to use guises, sorcery, and supernatural powers to appear and disappear at will."[92] In the words of Adam Malik, in the mind of the general public in 1940s, it was not impossible for Tan Malaka to have these magical abilities. He "was rumored to have the miraculous ability to disappear at will and to appear in two various places at the same time. Such fantastic rumors are what turned him into a superhuman being on an equal footing with the gods."[93] The equation of a revolutionary with the gods is not surprising. The labors of a revolutionary have often been depicted as "immense, gigantic, and intercontinental," and they seem to demonstrate "an enormous transition in human history."[94] Perhaps Tan Malaka's ability to cross geographical, cultural, and linguistic boundaries and to move avidly along transnational and transcultural terrains as he organized an anticolonial movement is so enormous that it only seems to make sense that he had godly abilities. However, a momentous historical transition, including the endeavor to organize and mobilize anticolonial struggles in an international solidarity beyond the boundaries of a colony could not be the making of a few people; in contrast, as this chapter shows, it involved the work of many ordinary sailors in their daily setting affecting their mundane lives in multinational ports and aboard the ships.

Even though the *pergerakan merah* receded from the interior of Java, Sumatra, and other islands, it lived on, however surreptitiously, in the shipping lanes and port harbors, surviving to fight another day. Through clandestine activities of writing and reading, publishing, translating, and smuggling letters and books, sailors and port workers continued igniting the fire of red enlightenment. Just as the *pergerakan merah* and its red enlightenment came to be in the Indies through communication networks, notably shipping lanes and ports, in the aftermath of their banning they survived in these circuitries.

CODA

The Enlightenment Project Is Necessary and Unending

In official archives, ordinary people appeared in fragments. Rarely chronicled in a full view, they occupied pages of reports and mainstream newspapers in number, or—out of the spotlight—as a background for surveilled main leaders. In the aftermath of the 1926 to 1927 revolts, Dutch papers reported the list of lesser-known names who had been arrested and exiled to the new internment camp at Digoel. Among them were *woro* Moenasiah (thirty), *woro* Ati (twenty-five), *woro* Soetitah (thirty-seven), Prapto (thirty), and Soemantri (thirty-four). Dja-maloedin Tamin joined them a few years later, as did our sailor students, Kamoe, Wentoek, and Waworoentoe. In these papers, the fate of our one and only female editor, *woro* Djoeinah, however, remains a mystery—her name did not make the list.[1] Such is the picture of ordinary people in formal documents. Relegated to a status of public nuisance, their names become pieces of a puzzle that never fully come together.

The story that emerges from this book is a story of ordinary lower-class people actively and creatively producing their own revolutionary communication to propel the *pergerakan merah* in 1920s. By exploring *openbare vergaderingen*, People's Schools, the *pers revolusionair*, and communication workers, we are able to witness the magnitude of the *pergerakan merah*. Though we do not know all their names, we are able to intimately delve into their feelings of suffering, expressions of hope, and imagination of a different future. The context that shaped this movement, however, went back to the Enlightenment era. Much like in the metropole, in the long 1800s through to the beginning of the 1900s, Enlightenment ideas shaped Dutch colonial policy and its contours of communication networks in the Indies, changing the class structure of the natives from peasantry to proletarian and expanding their modes of mobility and sociability. The suffering under colonialism and feudalism, coupled with access to print matters and transnational communication networks, allowed for emancipatory rhetoric and language of enlightenment to spread, first, among the elite natives, and, later, to the *kromo* class. Within the backdrop of the roarin' twenties, which was characterized by global cultural exchanges and interaction, *kromo* people adopted and repurposed existing emancipatory language of enlightenment and combined it with novel radical ideas of communism, creating a red enlightenment that became the basic aspiration propelling the movement. As such, the movement's day-to-day organizing process revolved around the democratic production of *berkeroekoenan* (a solidarity-based organization), meetings or assemblies, the arts, and activities around education, including reading, writing, publishing, and schooling—creating not just the politics of resistance but more importantly cultures of resistance. Hence, red enlightenment in this period became a unique event of mediation that shaped their strategies of mobilization and socialization. Consistent throughout the movement both in its above- and underground forms, these activities promoted resistance against capitalist exploitation and colonial repression in the name of *kemanoesiaan* (humanity). Hence, the movement was global and cosmopolitan in its network and vision, proposing a liberation of the Indies from colonial rule as part of the larger project of emancipation of international workers and the oppressed from colonialism and capitalism.

It is, therefore, apt to call this period the "*pergerakan merah*," as the revolutionaries called it, instead of the "*pergerakan*," as it is recognized in the existing literature. The *pergerakan* era is often associated with a period of National Awakening when elite *prijaji* adopted modern education and organization with the goal of achieving progress for natives that they identified as either Boemi Poetera (natives) or Muslims. Scholars often describe the communist period of the 1920s as a chapter within the *pergerakan* era when native awakening included popular mobilization of lower-class people.[2] However, this book demonstrates that the marker "*merah*" (red) is a necessary allegory for vital characteristics that are otherwise not captured by the term *pergerakan*. First, the *pergerakan merah* was united by an identity of *kromo* (later proletariat). It was a conscious self-creation of a collective identity that was based on their material economic condition. It differentiated them from the class of the elites and the capitalists, regardless of ethnic, national, and religious backgrounds. Second, they were able to see themselves as a part of a global and transnational community of international workers and the oppressed. Before the National Revolution decades later, the anticolonial movement was mobilized by the *kromo* in the Indies with a vision to free *all* international workers from their chains; national liberation was only a part of their goal.[3] Third, the consciousness of the *kromo* class was forged and reproduced through the production of its own revolutionary communication. The *kromo* borrowed, adapted, translated, transliterated, and repurposed transnational sources of arts, ideas, technologies, and strategies to express their suffering in order to envision their collective hope for humanity. The expression and recognition of suffering, as expressed in meetings and writings, was key to solidarity and movement building. A vision of collective hope and an alternative future was thus created within the revolutionary communication they produced. This massive movement of intellectual energy and emancipatory potential was destroyed following the banning of communism in 1926 to 1927.

Why does the history of the *pergerakan merah* matter *for us*? Beyond Indonesia, this book highlights the value of understanding the ebb and flow of enlightenment ideas. In an attempt to argue for the discontinuity between "Enlightenment and what follows," we might have been "overlooking some form of continuity," as Clifford Siskin also insists.[4] This is not just a task of expanding our account of enlightenment to include the "Radical Enlightenment" that looks past *les grands philosophes* of the French tradition and instead celebrates the subterranean and transcontinental networks of obscure thinkers and publishers.[5] It is also an endeavor to seek evidence of continuity of the Enlightenment project beyond Europe in the eighteenth century. Today, a number of scholarly works demonstrates this continuity, including in the history of the late eighteenth to mid-twentieth century women of letters in France, of enlightenment in nineteenth-century Japan and the Philippines, as well as the May Fourth movement in twentieth-century China.[6] Reflecting on the experience of the *pergerakan merah*, I argue that the enlightenment project and the revolutions it animates offer a historical legacy that is worth saving, including the creation of a radical space of universality that is characterized by democratic participation and equality of access, collective senses of belonging, and a progressive sense of social justice.

This book is founded on the premise that the enlightenment project is global and ordinary, and a product of continuous struggle. The vision of human emancipation could not be universal if it stopped in the eighteenth century, if it was geographically limited only to Europe and the Atlantic world, if it was only produced by a few male philosophers or revolutionary leaders, and if it only consisted of a set of clichés and moralistic slogans. Beyond the question of origin, this book unearths the ordinary processes that sustain the global enlightenment project as a common struggle for universal ends in the name of humanity and attends to the changing contradictions and conditions that necessitate such struggle. Universalism here should not be understood as Eurocentric; instead, as this book shows, universalism means discovering commonalities. Within academia, Enlightenment's Eurocentric form of universalism, like the one that motivates colonialism, is often responded to with a turn to difference.[7] In other words, universal ideas are rejected as differences are celebrated.[8] Through this lens, we fall short by saying that resistance in non-western societies against colonialism was triggered only by their identity as "non-Western," and by not being able to view resisters as complex beings who share fundamental values and aspirations with their fellow humans around the world.

The *pergerakan merah*, for example, is either framed as a national struggle or seen as a parochial endeavor that is only relevant for Indonesian people or for postcolonial societies. Under this narrative, new language and new communicative technology produced by the movement were all subsumed under the expedient narrative of national struggle. Anticolonial struggles were narrowly understood as struggles between the Dutch and the Indonesian natives. However, as this book shows, the resistance against Dutch rule was waged by people with multiple racial backgrounds, be they Indonesian, Dutch, or Chinese. On the other side, some natives supported and advocated prolonging Dutch colonialism. Attending to difference without accounting for universal characteristics, like our collective need to live freely with equal dignity and rights, prevents us from evaluating oppression and exploitation through a shared human experience. This antinomy of universalism versus difference forecloses our ability to see similar impulses between resistance movements across the globe, from the Haitian revolution to the Paris Commune to the US Civil Rights movements to anticolonial movements worldwide. All these movements struggled for the liberation from the dehumanizing effects of slavery, feudalism, colonialism, and imperialism. As such, the antipodes of colonialism—and more generally capitalism— are, as this book shows, *common struggles* for universal emancipation. In other words, the antithesis of the Eurocentric form of universalism is not difference, but rather our common suffering under capitalism—with its diversity of manifestations that we experience—and hence of our collective hope for a more just and humanizing world.

Thinking about enlightenment dialectically as a project of struggle allows us to understand that its universal character is found in a continuous process of becoming, and it comes into being and becomes universalized in moments of resistance against the equally universalizing system of capitalism and imperialism that necessitates such struggles. In other words, the systemic structures of exploitation and alienation that are inherent in capitalism and imperialism become the condition that produce demands for universal emancipation. Additionally, in order to be truly universal, emancipation needs to be fought for by people across racial, linguistic, gender, class, and national backgrounds. Haitian slaves, Black Americans, the colonized, the poor, workers, and the subalterns from around the globe made the enlightenment universal by taking up its ideals of human

rights, liberation, and social justice and turning them against Western colonizers, imperialists, capitalists, and their supporters. Democratic social movements,
therefore, become the location of the continuation of the enlightenment project.
Djoeinah and the Indonesian revolutionaries should then be seen as a part of this
long revolution that we are still collectively mobilizing today.[9]

Borrowing from the historian Sebastian Conrad, I argue that the enlightenment project was not simply a moment of the past that was bound to a specific
geographical space. It is, rather, an ensemble of continued processes of global
circulation, translation, entanglements, and transnational coproduction involving ordinary people of different sexes, races, nationalities, religions, and ages. The
enlightenment project cannot be read merely as a product of the West; rather, it is
the product of a complex amalgamation of tactics and inspirations coming from
other movements across the globe. In moments of contact between circulating
ideas and their translation, reinterpretation, and rearticulation, universal ideas of
solidarity emerge. Seeing the enlightenment project this way helps us understand
that the universality of enlightenment was the work of many different people in
the West and the East, South, and North, manifested through diverse cultural
contacts and conflicts, and often conditioned by unequal power relations as seen
in colonialism, imperialism, and racism.[10] From this perspective, the enlightenment project is a project of struggle. The question, then, is not so much about the
origin of enlightenment, but rather how to create spaces for universal impulses
and concerns, regarding social justice, equality, liberation, and emancipation in
plural and diverse expressions and manifestations in different parts of the world
and in different cultural contexts, and to identify the conditions that are able to
revive and sustain them in service of radical politics for social justice.

This book shows that revolutionary communication is one such space and
place for an enlightenment project. Its practice offers a key historical site for
transnational mobilization and organization to demand universal emancipation.
By attending to the complexities of the production of revolutionary communication as a space to assess the global history of enlightenment, this book unearths
sources of democratic communication that are rooted in it. It demonstrates that
existing struggles for social change and social justice will require the inclusion of
radical changes in the way our global sources of communication are organized
and managed. Revolutionary communication manifested as both democratic
means and ends for the struggle of universal emancipation. As it endeavored to
overthrow exploitation, suppression, and repression under colonialism, it contained a vision of how communication operated in the new desired society: egalitarian, inclusive, participatory, and open to diverse expressions and opinions.

The global, contentious, and ordinary nature of the enlightenment project as
a social movement shapes the character of revolutionary communication. The

first characteristic is that revolutionary communication, like other democratic communication traditions, is a site of exchanges, borrowing, translation, transformation, and adoption of communicative repertoires of contention. Universal emancipation emerges and the global enlightenment project continues to evolve as communicative practices of resistance and methods of resistance traveled around the world and were borrowed, adopted, and repurposed in different time and space. This act of borrowing and transformation is particularly vivid in the democratic revolutions that the enlightenment animated and the democratic communication the movements produced.

The next characteristic of revolutionary communication is the act of creative repurposing. While the existing colonial communication infrastructures were intended to expand colonial power, the people in the movement creatively repurposed and brought them together into "revolutionary communication," revealing the centrality of media in the making of a social enterprise. These creative repurposing practices point to the nature of communicative sociotechnical systems as projects and sites of struggle. To think of an enlightenment project as a contentious practice means to understand assembly, association, literacy, and the public sphere as forms of "popular contentions." Thinking of the public sphere as a Habermasian bourgeois public sphere, for example, which excluded the poor and women, differs from the revolutionary and plebeian public spheres and the Bengali *adda* in their democratizing nature.[11]

Revolutionary communication also demonstrates the immanent character of the enlightenment, that is, to draw power not from something external to the society, but to take up the existing means of struggle and repurpose them into ones of resistance. In Karl Marx's words, "[t]he bourgeoisie itself, therefore, supplies the proletariat with its own elements of political and general education, in other words, it furnishes the proletariat with weapons for fighting the bourgeoisie."[12] Communication work is revolutionary work and is indispensable in building and organizing a movement. It is equally indispensable to imagining a new world.

Last, the ordinary character of an enlightenment project means that revolutionary communication was produced by ordinary people through ordinary processes and actions as they organized themselves in the anticolonial and antiimperialist movements. The basis of a democratic system is that everyday people should have control over political, economic, and social resources, and the control over the means of communication is an integral aspect of such power.

The late professor of journalism Todd Gitlin once wrote that "the enlightenment project is necessary and unending."[13] In mobilizing its red enlightenment, the history of the *pergerakan merah* shows that, despite the oppressive and exploitative colonial environment, lower-class *kromo* people united in solidarity risking much for a better world. The collective hope, based on their myriad experiences of suffering, allowed them to imagine a different world and to struggle for it. *Rasa* (feeling) became the basis of the connection and solidarity, key to their mobilization. By way of the revolutionary communication, the *kromo* found self-expression of their collective identity. As language creates consciousness, media creates community and organization. Despite the demise of the movement, this emancipatory spirit survived in communication circuitries, media, language, and cultures of resistance ready to be revived and repurposed by the next movements. The *pergerakan merah* serves as a reminder that in the extreme suffering that capitalism imposes on us, there is hope that another world is possible.

LIST OF CARTOONS PUBLISHED IN *SINAR HINDIA/API* IN 1918–26

The following are listed in order of publication date. Included is the text that accompanied each cartoon.

Tiger and water buffalo (*Sinar Hindia*, January 12, 1924)

(See figure 1 in chapter 2.)

> *"Kalau kaoem boeroeh bersatoe dan revolutionair, tidak sadja ia bisa mendapat nasib baik, akan tetapi ia bisa mengoesir kapitalisme, sebagai kerbau di atas jang menandoek harimau dan babi hoetan. Karena itoe, kaoem boeroeh bersatoelah."*

> "If workers are united and revolutionary, not only could they have a good luck, but they could also kick out capitalism like a water buffalo that headbutts a tiger and a wild boar. Therefore, workers, unite!"

Monkey and water buffalo (*Sinar Hindia*, January 12, 1924)

(See figure 2 in chapter 2.)

> *"Kalau kaoem boeroeh terlaloe sabar, kaoem kapital djadi koerangadjar, sebagai mojet [sic, monjet] jang mengentjingi kepalanja kerbau diatas ini."*

> "If workers are too patient, the capitalists will become insolent just like a monkey that urinates on the head of a buffalo."

Thrift, theft (*Sinar Hindia*, April 2, 1924)

Penghematan.   ‖   Pentjoerian.

Hajo lekas bajar !
Dan kamoe pegawai   ketjil-ketjil boleh pergi !
Petibesi tidak boleh kosong !

"*Penghematan. Pentjoerian.*"
"*Hayo lekas bajar!*
Dan kamoe pegawai ketjil-ketjil boleh pergi!
Petibesi tidak boleh kosong!"

"Thrift. Theft."
"Come on, pay quickly!
And you little employees can go!
The safe cannot be empty!"

Crocodiles created a *vergadering* (*Sinar Hindia*, June 6, 1924)

(See figure 3 in chapter 2.)

"*Boeaja memboeat vergadering.*"
"*Gambar di atas ini meloekiskan bagaimana boeaja-boeaja jang soedah*
kehilangan akal oentoek mentjari pengaroeh di moeka golongannja
hingga riboet sekali pikirannja.

Dengan diam-diam sesoedah ia memboeka kongresnja di Garoet dimana
ia memboeat roepa-roepa voordracht laloe teroes ke Bandoeng oen-
toek memboeka poela vergadering jang kedoea kalinja.
Barang siapa bisa menebak teka teki itoe, siapakah boeaja itoe akan
dapat Sinar satoe minggoe vrij.
Kalau tebakan terlaloe banjak, akan di oendi."

"Crocodiles created a *vergadering*."

"This picture above illustrates how crocodiles have lost their minds to
influence their people that it created a chaos in their own heads.

On the sly, after he opened the congress in Garoet where he created
various *voordracht* [Dutch: speeches], he then went on to Bandoeng
to open a second *vergadering*.

Those who can guess who the crocodile is can receive *Sinar* for a week
for free.

If there are too many guesses, [winner] will be drawn."

Workers who are squeezed like laundry (*Sinar Hindia*, June 21, 1924)

Kaoem boeroeh jang selagi diperas sebagai
kain jang habis ditjoeji.

"Kaoem boeroeh jang selagi diperas sebagai kain jang habis ditjoetji."
"Demikianlah kaoem kerdja selaloe mendjadi korban, selama ia masih
terpitjah-pitjah oleh kebangsaan dan keagamaan!

Ia tidak akan berenti diperas sebeloem kapitalisme hantjoer dan Kom-munisme lahir.

Oleh karena itoe: PROLETAR SELOEROEH DOENIA, DARI SEG-ALA BANGSA DAN AGAMA BERSATOELAH!"

"The workers are currently being squeezed like cloth that has just been washed."

"The working class always becomes a victim, as long as it is still divided by nationality and religion!

They will not stop being squeezed before capitalism is destroyed and communism is born.

Therefore: *PROLETARIATS OF THE WORLD FROM ALL NATIONS AND RELIGIONS UNITE!"*

Betawi-Manilla whispering (*Sinar Hindia*, July 5, 1924)

BETAWI – MANILLA BISIK- BISIK

"BETAWI-MANILLA BISIK-BISIK"
"Hoedjan rintik-rintik
Katjang diperahoe
Sana bisik-bisik
Ra'jat soedah tahoe."

"BETAWI-MANILLA WHISPERING"
"Drizzling
Peanuts in a boat
Whisper there
The people already know."

Japan–America (*Sinar Hindia*, July 12, 1924)

─DJEPANG ─ AMERIKA─

"Djepang—Amerika"

"Amerika: *Hajo, Djepang kau maoe rewel? Boleh kita adoe djotosan!*

Djepang kepada Paus: *Datoek, itoe bagaimana Amerika begitoe ganas. Saya dipelotot-pelototi sadja diadjak berkelai.*

Paus: *Ja, berkelai itoe memang dilarang oleh agama. Tetapi bagaimana lagi itoe agama soedah lama disimpan dalam* [not clear] *kaoem modal dari segala bangsa."*

"JAPAN—AMERICA"

"AMERICA: Come on, Japan, are you going to be fussy? We can have a fistfight!

JAPAN TO THE POPE: Grandfather, that's why America is so vicious. I've been glared at and invited to fight.

THE POPE: Yes, fighting is prohibited by religion. But how else[?] Religion has long been kept in [not clear] of the capitalists of all nations."

Islam stolen by Islam (*Sinar Hindia*, July 14, 1924; republished July 15, 1924)

————TEKA – TEKI————

ISLAM DITJOERI ISLAM!

HANTOE MERAH MEMEGANG PALOE :
„Maoe kau bawa kemana oeang itoe?"

PENIPOE: „Ach toean, mengapa toean marah? Toh soedah selajaknja oeang Islam ditjoeri oleh bangsat jang berselimoet Islam!"

"*TEKA-TEKI*"

"*ISLAM DITJOERI ISLAM!*"

"HANTOE MERAH MEMEGANG PALOE: '*Maoe kau bawa kemana oeang itoe?*'

PENIPOE: '*Ach toean, mengapa toean marah? Toh soedah selajaknja oeang Islam ditjoeri oleh bangsat jang berselimoet Islam!*'"

"A PUZZLE"

"ISLAM STOLEN BY ISLAM!"

"A Red Specter Held a Hammer: 'Where are you taking that money to?'

Thief: 'Oh, sir, why are you angry? Isn't it appropriate for Islamic money [Muslims to have their money] be stolen by a bastard covered [blanketed] in Islam[?]'"

"Rawe-rawe rantas, malang-malang poetoeng" (Sinar Hindia, July 19, 1924)

RAWE-RAWE RANTAS, MALANG-MALANG POETOENG.

"*RAWE-RAWE RANTAS, MALANG-MALANG POETOENG*"

"*Lihatlah itoe Ra'jat, kaoem-kerdja international menggoelingkan doenia kapitalisme!*

Meskipoen kaoem kapital dan sekalian jang reactionair menghalang-halangi, tetapi . . . . . . lihat itoe! . . . . . . mereka mesti tergilas dan di indjak-indjak.

Demikianlah toekang gambar kita meibaratkan perdjoeangan kaoem atau klassenstrijd itoe bagoes sekali!"

"RAWE-RAWE RANTAS, MALANG-MALANG POETOENG"

[An Indonesian proverb: lit. "Plants that stretch out must be cut down completely and those that block the road must be broken"; my translation: "Everything that hinders the aims and objectives must be removed."]

"Look at the people, the international workers, overthrowing the world of capitalism!

Even though the capitalists and all the reactionaries are obstructing it, but . . . look at that! . . . they must be crushed and trampled.

That's how our cartoonist depicts how good the struggle of the people or class struggle is!"

You want it or not (*Sinar Hindia*, July 26, 1924)

MAOE ATAUPOEN TIDAK.

"*Maoe ataupoen tidak Ra'jat-Ra'jat disekalian negeri kemodalan mesti tertarik masoek djoerang dari beratnja padjak-padjak jang djoega dikorbankan kepada 'Dewa-Perang.'*

Walaupoen begitoe dari djoerang kemiskinan mereka melihat matahari *sociale revolutie jang di hiasi oleh sinar* paloe arit *makin lama makin tinggi. Dan matahari itoelah nanti akan membinasakan* Dewa Modal dan Dewa perang *achirnja mendatangkan Kommunisme!*"

"Whether we want it or not, people in all capitalist countries must be
drawn into the abyss of the heavy taxes that are also sacrificed to
the 'God of War.'

Even so, from the abyss of poverty they saw *the sun of social revolution*
decorated with the rays of *hammer and sickle* getting higher and
higher. And that sun will later destroy *the God of Capital and the
God of War* and finally bring about Communism!"

B. O. in Volksraad (*Sinar Hindia*, August 2, 1924)

(See figure 4 in chapter 2.)

"*B.O. dalam Volksraad*"
"*Kandjeng toean voorzitter jang terhormat!*
Betoel saja seorang Djawa, tetapi saja tahoe adat dan sopan santoen"

"B. O. in Volksraad"
"Dear Mr. Chairman!
It is true that I am a Javanese, but I know customs and manners."

After medical workers' strike (*API*, August 18, 1925)

SETELAH PEMOGOKAN DJOEROE DJOEROE RAWAT.

"Setelah pemogokan djoeroe djoeroe rawat."

*"75 pCt pegawai sama mogok, karena beratnja pekerdja'an serta ban-
jaknja sewenang-wenang.*

75 pCt pegawai sama mogok, karena menoentoet perbaikan nasib.

*Pendjilat bertambah berat pekerdja'annja, karena dapat extra
pekerdja'annja pemogok-pemogok.*

*Pekerdja'an soedah mendjadi berat, tetapi madjikan saban hari marah-
marah seperti matjan."*

"After Medical Workers' Strike."

"75 percent of employees are on strike, because of how hard the work
is and how much arbitrariness there is.

75 percent of employees are on strike, because they demand improve-
ments in their fate.

The job of a sycophant is harder, because they get extra work that the
strikers have left.

The work has become hard, but the boss is angry every day like a tiger."

After the strike in a hospital (*API*, August 19, 1925)

Setelah pemogokan di Roemahsakit.

"Setelah pemogokan di roemahsakit"
2 orang pendjilat P dan K

P. *Wah soesah benar badankoe ini, kerdja soedah berat lagi, soesah gendong njonjah jang djatoeh pingsan.*

K. *Lo, kena apa njonjah djatoeh pingsan?*

P. *Ja, sebab leerling-leerling semoea sama mogok.*

K. *O, barangkali dari meratapnja R., hingga njonjah dapat halangan pada zenuwennja.*

P. *Ja, boleh djadi.*

K. *Kemana larinja anak-anak itoe?*

P. *O, pintar sekali anak-anak itoe, mereka lari ke belakang dengan tidak diketahoei oleh orang.*

K. *Bagaimana kami ini? Bekerdja begitoe berat, teman tidak ada, poelang tak bisa merdeka . . . berangkat ke kantor dan poelang ke roemah moesti diangkat dengan auto, kalau kerdja ada lambat sedikit digetak getak . . . bagaimana toh ini?*

P. *La, habis bagaimana? Kalau saja, lebih baik kerdja teroes sadja . . . sebab sajang . . . pensioenmoe K?*

K. *Ja soedah, apa boleh boeat biarpoen itoe pensioen masih toenggoe 25 tahoen lagi dan biar kita dapat merk lidah pandjang, asal sadja dapat moeka manis dari madjikan, itoe soedah tjoekoep."*

"After the strike in a hospital"
Two sycophants P and K

P. Wow, my body is really having a hard time, work is really hard, it's hard to carry madam who has fainted.

K. O, why did madam faint?

P. Yeah, because the student nurses are all on strike.

K. O, perhaps due to R's wailing, madam got a nerve blockage.

Q. Yes, maybe.

K. Where did the students run?

P. O, they are so smart, they ran behind without anyone knowing.

K. How about us? Working so hard, no friends, we can't go home freely . . . going to the office and coming home you have to be picked up by a car, if work is slow, we get yelled at . . . So?

P. Well, so what? For me, it's better to just keep working . . . because [let's not waste] . . . your retirement, K?

K. Yeah what can we do? Even though the retirement is still another twenty-five years and we are marked long tongue [sycophant], as long as we get a sweet face from our employer, that's enough."

After the strike in Stadsverband (*API*, August 20, 1925)

This is the same image that was used in the August 18 edition, above (titled "After medical workers' strike"), but with this different title and the following text:

> *"Setelah pemogokan di stadverband"*
> *Toean Blondo seorang chef.*
> *Mentrei Troeno pendjilat.*
> *Koeldi leerling jang baroe sakit.*

> TROENO: *Wah soesah amat Koeldi . . . teman-teman sama mogok, sekarang saja jang moesti kerdja berat. Mantri kok disoeroe kerdja* [not clear] *toh Di?*
>
> Koeldi: *Itoe salahmoe sendiri Troeno. Karena kamoe soedah salah, ja rasakan sendiri nasibmoe. Sedang saja jang sakit ada soesah, sedikit soesah karena pada esoek harinja saja tidak dapat sarapan, senangnja, saja ta' disoeroeh kerdja seperti kamoe.*
>
> BLONDO: *Djangan tjrewet ja . . . kamoe moesti bekerdja keras . . . beloem boleh poelang, kalau pekerdja'an beloen beres. Kalau kamoe rewel-rewel, nanti kamoe saja ontslag sama sekali.*
>
> TROENO: *Tidak toean . . . Biarpoen kerdja berat sampai setengah mampoes saja maoe asal sadja tidak diontslag."*

"After the Strike in Stadsverband"
Mr. Blondo [Dutch] is a chief.
Head Nurse Troeno is a sycophant.
Koeldi is a student nurse who just got sick.

> TROENO: Wow, it's really difficult, Koeldi . . . my friends are on strike, now I have to work hard. How a head nurse is told to work [not clear]?
>
> KOELDI: It's your own fault, Troeno. Because you made a mistake, feel your own fate. When I was sick, it was difficult, a little difficult because the next day I couldn't have breakfast. The good thing is, I am not asked to work like you.
>
> BLONDO: Don't be so fussy . . . you have to work hard . . . you can't go home yet, if the work isn't finished. If you're fussy, I'll fire you.
>
> TROENO: No sir. . . . Even if I have to work hard until I'm half dead, I'm willing as long as I am not fired."

Appendix B

LIST OF SONGS PUBLISHED IN *SINAR HINDIA/API* IN 1918–26

MARSCH SOCIALIST	MARSCH SOCIALIST (SOCIALIST MARCH)
Lagoe: Socialisieke [sic!] marsch	Song: Socialisieke [sic!] marsch
Soewardi Soerjaningrat	Soewardi Soerjaningrat
1.	1.
Hai, bangsa socialist, sadarlah!	Hey, socialist nation, wake up!
Ikoetkan bandera merah!	Follow the red flags!
Goena merdikakanlah kerdja!	To free up work!
Dari koengkoengan wang harta!	From the confines of money!
Keslamatan, kenikmatan,	Safety, enjoyment,
Trang fikiran, pengatahoean.	Bright mind, knowledge.
Haroes terpegang kaoem kerdja,	Must be held by the working people,
Wadjiblah kita mengedjarnja!	We must pursue it!
Itoe maksoed kita jang soetji moelia!	That is our holy, noble intention!
(2 kali)	(2 times)
Haroes menang, ra'jat kita! (2 kali)	Must win, our people! (2 times)
Refrien	*Refrain*

(Continued)

(Continued)

MARSCH SOCIALIST	MARSCH SOCIALIST (SOCIALIST MARCH)
2.	2.
Hai, kamoe jang berdjoeang njata!	Hi, you who are fighting for real!
Dimana tempat poen djoega!	Anywhere!
Bekerdja hingga soesah pajah	Work hard
Poen gadjih amat terendah.	Also the salary is very low.
Sigralah datang berikat!	Come and bond immediately!
Persatoean mendjadi koeat!	Unity becomes strong!
Terlepas dari masih ba'ja!	Regardless of being old!
Wadjiblah kita mengedjarnja!	We must pursue it!
Refrien	*Refrain*
3.	3.
Kita prang ta' dengan sendjata	We don't fight with weapons
Palroon/Palroeu, mimis, pisau wadja (?)	*Palroon/Palroeu, mimis, wadja* knife
Tadjamnja sendjata fikiran	The sharpness of the mind as a weapon
Menangkan kita berlawan!	Will win us who fight!
Damai, makmoer, tambah senang	Peace, prosperity, more joy
Hidoepnja sekalian orang,	Everyone's life,
Moelianja nasib kaoem kerdja	The noble fate of the working people
Wadjiblah kita mengedjarnja!	We must go after!
Refrien	*Refrain*

Source: *Sinar Hindia*, May 5, 1920.

SAIR INTERNASIONAL	SAIR INTERNATIONAL (LYRIC OF INTERNATIONALE)
Lagoe: Internationale	Song: Internationale
Soewardi Soerjaningrat	Soewardi Soerjaningrat
1.	1.
Bangoenlah, bangsa jang terhina!	Wake up, humiliated nation!
Bangoenlah, kamoe jang lapar!	Wake up, you hungry ones!
Kehendak jang moelia dalam doenia	Noble will in the world
Senantiasa tambah besar.	Always getting bigger.
L'njaplah adat fikiran toea!	Gone are the old habits of thought!
Hamba-ra'jat, sadar, sadar!	Servants of the people, be aware, be aware!
Doenia telah berganti roepa,	The world has changed,
Nafsoelah soedah tersebar!	Lust has spread!
Kawan kawan, hai, ingatlah!)	Friends, hey, remember!)
Ajo, madjoe berperang!) 2 kali	Come on, go to war!) *2 times*
Serikat Internationale, jalah) Refrein	The Union of Internationale, is) *Refrain*
Pertalian orang!)	A people's tie!)

SAIR INTERNASIONAL	SAIR INTERNATIONAL (LYRIC OF INTERNATIONALE)
2.	2.
Negri tindas, hoekoem berdjoesta.	The land of oppression, the law is lying.
Jang kaja troes hidoep seneng.	The rich continue to live happily.
Orang miskin terisap darahnja;	The poor have their blood sucked;
Ta'sekali berhak orang	People have no right at all
Djangan soeka lagi terperintah!	Don't like being ordered anymore!
Ingat akan persamaan!	Remember equality!
Wadjib dan hak tiada berpisah.	Obligations and rights are not separate.
Hak wadjib haroes sepadan.	Rights obligations must be commensurate.
Kawan-kawan . . . enz.)	Friends . . . etc.)
Sarekat Inter . . . enz.) 2 kali refrein,	Sarekat Inter . . . etc.) 2 times Refrain,
3.	3.
Penindes berfikiran sjaitan;	The oppressor has satanic thoughts;
Selaloe meratjoen kita.	Always poisoning us.
Djangan bantoe lainnja kawan-kawan!	Don't help others, friends!
Hai, bersatoelah oesaha!	Hi, unite efforts!
Moesoeh kita mendidik pahlawan	Our enemies educate heroes
Dalam golongan kita.	In our group.
Kepada jang brani melawan	To those who dare to fight
Kita djatohken sendjata!	We drop our weapons!
Kawan-kawan . . . enz.)	Friends . . . etc.)
Sarekat Inter . . . enz.) 2 kali refrein	Sarekat Inter . . . etc.) 2 times Refrain

Source: Sinar Hindia, May 5, 1920.

*Another version of this song appears in the newspaper in 1921 with slightly different lyrics (see below).

SAIR INTERNASIONAL	SAIR INTERNATIONAL (LYRIC OF INTERNATIONALE)
1.	1.
Bangoenlah, kaoem jang terhina.	Wake up, humiliated people.
Bangoenlah, engkau jang lapar.	Wake up, you hungry one.
Kehendak jang moelia dalam doenia	Noble will in the world
Senantiasa tambah besar.	Always getting bigger.
L'njaplah adat serta faham toea.	Old traditions and understandings are gone.
Hamba-ra'jat, sadar, sadar.	Servants of the people, be aware, be aware.
Doenia telah berganti roepa,	The world has changed,
Nafsoe soedah tersebar.	Lust has spread.
Pertandingan penghabisan, koem)	Final match,)
Poellah berlawan.) 2 kali	Gather to resist.) 2 times
Serikat Internationale, misti di doenia)	The Union of Internationale, must be in the world.)

(Continued)

(Continued)

SAIR INTERNASIONAL	SAIR INTERNATIONAL (LYRIC OF INTERNATIONALE)
2.	2.
Harta menindas dan lagi berdjoesta.	Wealth oppresses and lies.
Jang kaja teroes hidoep seneng.	The rich continue to live happily.
Orang miskin terisap darahnja;	The poor have their blood sucked;
Djanganlah soeka lagi terperas,	Don't like to be squeezed any more,
Ingat akan persamaan,	Remember equality,
Wadjib dan hak tidak berpisah.	Obligations and rights are not separated.
Hak wadjib haroes sepadan.	Mandatory rights must be commensurate.
Pertandingan penghabisan,)	Final match,)
koempoellah berlawan.) 2 kali	Gather to resist.) *2 times*
Serikat Internationale, misti di doenia)	The Union of Internationale,)
	must be in the world.)
3.	3.
Penindas berfikiran sjatan;	The oppressor has satanic thoughts;
Selaloe meratjoen kita.	Always poisoning us.
Djangan bantoe laskarnja lain orang,	Don't help other people's troops,
Bantoelah laskar sendiri,	Help your own troops,
Moesoeh kita mendidik pahlawan	Our enemies educate heroes
Dalam golongan kita.	In our group.
Kepada jang brani melawan	To those who dare to fight
Kita djatohken sendjata.	We drop our weapons.
Pertandingan penghabisan,)	Final match,)
koempoellah berlawan.) 2 kali	Gather to resist.) *2 times*
Serikat Internationale, misti di doenia)	The Union of Internationale, must be in the world.)

Source: Sinar Hindia, November 29, 1921.

BARISAN MOEDA	BARISAN MOEDA (THE YOUNG LINE)
Di timoer matahari	In the east of the sun
Moelai bertjahja	Starting to glow
Bangoenlah dan berdiri	Get up and stand
Kawankoe semoea!	All my friends!
Marilah kita bersedia	Let us be ready
Mengatoer barisan kita!	Arranging our ranks!
Kita barisan moeda	We are the young line
Dari kaoem pekerdja	From the workers

BARISAN MOEDA	BARISAN MOEDA (THE YOUNG LINE)
Kita menoeloeng bapak	We help you, father
Melawan si boeas.	Against the beast.
Kita mendjaga anak	We look after children
Hingga ta diperas.	Until they aren't squeezed.
Kita bersama melawan,	We fight together,
Mentjari keselamatan.	Seek safety.
Bangoen barisan moeda,	Build a young line,
Dari kaoem pekerdja!!	From the workers!!
Kita reboet persamaan	Let's get equality
Dan kemenoesiaan,	and humanity,
Persaudaraan, keadilan	Brotherhood, justice
Poen kemerdikaan.	Even independence.
Djika semoea tersampai	If everything arrives
Baroelah kita berenti.	Then we stop.
Madjoe barisan moeda	Forward the young line
Dari kaoem pekerdja.	From the workers.
Bapak kita tertindas	Our father was oppressed
Si emak terisap	The mother was sucked
Kita anak terperas	We are the squeezed [exploited] children
Hidoep tidak tetap	Life is not fixed
Kaoem modal senang-senang	Capitalists are having fun
Hidoepnja ta koerang-koerang	Their lives are not scarce
Bangoen barisan moeda,	Build a young line,
Dari kaoem pekerdja!!	From the workers!!
Bapak, anak dan kita	Father, son and us
Berkoempoel bersatoe	Gather unite
Bersama-sama kita	Together we are
Berdjalan menoedjoe	Walking towards
Ke Kemerdikaan menoesia	To human Independence
Di seloeroehnja doenia	All over the world
Bangoen barisan moeda,	Build a young line,
Dari kaoem pekerdja!!	From the workers!!

Source: *Sinar Hindia*, April 28, 1923

219

SAIR KEMERDIKAAN	SAIR KEMERDIKAAN (LYRIC OF INDEPENDENCE)
Melawan, melawan.	Fight, fight.
Kita memang pahlawan.	We are heroes.
Melawan, melawan!	Fight, fight!
Kita banjak kawan.	We have lots of friends.
Madjoe-madjoe	Move forward
Dengan keberanian,	With courage,
Hingga semoea kamoe	Until all of you
Di Kemerdikaan.	In Independence.
Merdika-merdika	Independence
Itoelah hak menoesia.	That's a human right.
Merdika-merdika,	Independence,
Itoelah hak kita.	That's our right.
Kaoem Kromo,	*Kromo* people,
Madjoelah melawan!	Come forward and fight!
Lengkaplah dan madjoe	Complete it and move on
Ke Kemerdikaan!!!	To Independence!!!

Source: *Sinar Hindia*, April 28, 1923.

ENAM DJAM BEKERDJA	ENAM DJAM BEKERDJA (SIX HOURS OF WORK)

5 | 5 . 5 -4- . 5 6 . 5 | i . 5 |
E - nam djam, soe - dah sam - pai la – ma   Six hours, it's a long time

5 | 3 . 3 -2-. 3 6. 5 | 2. . |
E - nam djam, se – ti - ap ha – ri   Six hours, every day

5 | 5 . 5 -4- . 5 6. 5 | i . 5 |
E - nam djam, me – me - ras te – na – ga   Six hours, squeezing out energy

i | 7 . 7 6 . 6 2̇. 2̇ | 5 . . |
Boe - at men – tja ri re – ze – ki   To find fortune

5 | 7 . 6 5 . 4 3 . 2 | 6 . 5 |
Di ta - nah jang ber – ha - wa pa – nas   In hot climates

3 | 5 . 4 2 . 2 3 . 4 | 3 . . |
E - nam djam soeng – goeh - lah sam – pai   Six hours are really enough

5 | 7 . 6 5 . 4 3 . 2 | 6 . 5 |
Ti - dak soe - ka ki - ta di - pe – ras   Don't like us being squeezed

i — 7 . 7 6 . 6 2̇ . 2̇ | 5 . . —
Le - bih da - ri la - ma la – gi   Much longer

5 | i . . i | 6 . . 6 | 2̇ . 3̇ 2̇ i | 7 .
E - nam, e - nam, e - nam, djam se- ha – ri   Six, six, six, hours a day

. 5 | i . i i . i i . i | i . . i |
Re – boet - lah, hai, ka - oem ker - dja. Re –   Seize it, you workers.

3̇ . i 5 . 5 6 . 7 | i . . ||
Boet - lah e - nam djam ker - dja.   Seize six hours of work.

II

Kapital siloba dan tama'	The capitalists who are greedy

Kapital siloba dan tama' — The capitalists who are greedy
Senantiasa bermaksoed — Always mean it
Isap darah kita terbanjak — To suck the most of our blood
Menjoesoet isinja peroet. — Shrinking our stomach's content.
Wadjiblah itoe kita lawan. — We must fight against that.
Dengan persatoean kita. — With our unity.
Persatoean dalam fikiran — Unity in mind
Dan persatoean tenaga — And unity of energy
Enam, enam, enam djam sehari. — Six, six, six hours a day.
Reboetlah, hai kaoem kerdja, — Seize it, O workers,
Reboetlah enam djam kerdja. — Seize six hours of work.

III

Biarpoen kita soedah dapat — Even though we get it
Kerdja enam djam sehari — Working six hours a day
Tapi selamanja si bangsat — But forever the bastard
Bisa sadja isap lagi. — can suck us again.
Dari itoe haroeslah linjap — That is why it must disappear
Kemodalan dari boemi. — Capital from the earth.
Haroeslah kita dengan tjepat. — We have to be quick.
Mengganti pratoeran negri. — Changing state regulations.
Sowjet, sowjet, haroes kita kedjar. — Soviet, soviet, we must go after them.
Datangkanlah, hai, Proletar. — Come, O, Proletariat.
Bendera merah berkibar. — The red flag is flying.

Source: Sinar Hindia, April 26, 1924.

Other songs sung in meetings during the *pergerakan merah*:
- "Proletar" (Proletariat; source: *Sinar Hindia,* November 12, 1923)
- "Socialisme" (Socialism; source: *Sinar Hindia,* November 12, 1923)
- "Bendera merah" (Red Flag; source: *Sinar Hindia,* November 24, 1923)
- "Darah rajat" (People's Blood; source: *Sinar Hindia,* April 28, 1924)
- "Meiviering" (Dutch: May Celebration; source: *Sinar Hindia,* May 9, 1924)
- "Perlawanan" (Resistance; source: *Api,* March 23, 1925)
- "Hidjo-hidjo" (Green-Green; source: *Api,* May 19, 1925)
- "Roode Garde" (Dutch: Red Guard; source: *Api,* June 4, 1925)

LIST OF SLOGANS PUBLISHED IN
SINAR HINDIA/API IN 1918–26

May 2, 1918

*"Hindia berdarah, Hindia berapi. Gasaklah jang salah. Dengan berani sampai
    mati!"*

"Bloody Indies, fiery Indies. Strike the wrong ones. Bravely unto death!"

February 20, 1922

"Tolonglah kaoem pemogok di pegadean!"

"Help the strikers at the pawn shop!"

February 25, 1922

"Ra'jat Hindia, batjalah SINAR-HINDIA! Awas, Kemerdika'an Kapitalisme."

"People of the Indies, read *SINAR-INDIA*! Beware, the Independence of
    Capitalism."

February 28, 1922

"Sinar Hindia akan memberi penerangan oentoek memboeka pikiran baroe."

"*Sinar Hindia* will give enlightenment to begin a new thought"

February 23, 1923

"Awas! Kalau maoe sehat fikiranmoe batjalah SINAR-HINDIA!"

"Watch out! If you want to keep your mind healthy, read *SINAR-HINDIA*!"

"Sokonglah SINAR moesoeh kapitalisme!"

"Support *SINAR*, the enemy of capitalism!"

May 8, 1923

"Kalau mogok tidak oesah angkat sendjata. Tidoer sadja diroemah, tentoe simadjikan peroetnja kempet."

"When on a strike, don't use weapons. Just stay at home. Truly your employer will have a flat stomach."

"Kaoem koemunis senantiasa bersedia oentoek membajar tindasannja si djahanam! Awaslah."

"The communists are always ready to pay back the devils' [capitalists'] exploitation! Beware."

"Hai kaoem proletar! Atoerlah dengan sigera barisan kita. Lemparlah si chianat kapitaalisten."

"O, proletariat! Arrange our ranks immediately. Throw down the capitalist traitor."

"Pemogokan Spoor dan Tram mesti djadi, mana kala permintaan tidak ditoeroeti dan gadjih ditoeroenkan, atau poen pemimpin diboeang. Awaslah, hai kawan2 [kawan-kawan] akan signaal pemogokan."

"Spoor and Tram strikes must occur, when demands are not complied with and salaries are reduced, or leaders are thrown out. Watch out, friends, for the strike signal."

"Djanganlah soeka menerima perkataan manis dari pihak sana; itoe semoea ratjoen! Pertjaja sadjalah atas toean poenja kekoeatan sendiri."

"Don't accept sweet words from the capitalists; they are all poisons! Believe in your own strength."

"Ketahuilah, apabila Spoor datangnja telaat, itoelah ertinja pemogokan moelai."

"You know, if Spoor arrives late, that means the strike has started."

"Kaoem boeroeh Hindia! Awaslah benar benar. Kawanmoe di spoor dan tram akan mogok, sebab."

"Indies workers! Be really careful. Your friend [working in] trains and trams will strike, because."

"Kaoem boeroeh Spoor dan Tram! Ingatlah betoel2 [betoel-betoel] tentang kesoekaran dalam toean-toean poenja lingkoengan."

"Spoor and Tram workers! Remember very well the difficulties in your own environment."

"Pemoeda Hindia! Djanganlah mendjadi pengetjoet pemogokan. Pimpinlah bangsamoe jang tertindas."

"Youth of the Indies! Don't be a strike coward. Lead your oppressed nation."

"*Kaoem Spoor-an! Mogok sadjalah kapan pemimpinmoe ditangkap! Apa bila toean toean tinggal diam, tjelakalah nasib toean toean dibelakang hari.*"

"The Spoor people! Just strike when your leader is arrested! If you gentlemen remain silent, woe to your fate in the future."

"*Djanganlah bersandar kepada takdir. Pikiran jang demikian itoe soedah kolot. Perkataan 'takdir' itoe tjoema timboel dari moeloetnja reactie.*"

"Don't rely on fate. That is an old idea. 'Fate' only comes from a reactionary's mouth."

February 18, 1924
"*Awas! Moelai 7 Maart d.m. SINAR akan terbit tiap-tiap hari Djoema'at djoega.*"
"Watch out! Starting March 7 *SINAR* will be published every Friday too."

"*SINAR Moesoeh kapitalisme, tetapi kawan Proletar,*"
"*SINAR*, enemy of capitalism, but friend of the proletariat,"

March 3, 1924
"*SINAR misti madjoe, oesahakanlah tambahnja lengganan.*"
"*SINAR* must progress, try to increase subscriptions."

March 6, 1924
"*Setia pada SINAR, ertinja menegoehkan kemenoesia'an.*"
"Being loyal to *SINAR* means defending humanity."

"*SINAR memang keras, tapi manis.*"
"*SINAR* is fierce, but also sweet."

March 7, 1924
"*Menoenggak pembajaran SINAR, ertinja djadi rem datangnja socialisme.*"
"Putting *SINAR* payments in arrears means it will be a brake on the arrival of socialism."

March 15, 1924
"*SINAR, itoelah obat otak boeroeh!*"
"*SINAR*, that's the medicine for your brain!"

March 17, 1924
"*SINAR, melawan kapitaal dan boedak-boedaknja.*"
"*SINAR* fights against capitalism and its slaves."

March 18, 1924
"*SINAR, memberitakan pergontjangan boeroeh.*"
"*SINAR*, reporting on labor unrest."

"*SINAR, adalah tempat pikiran Boeroeh*"
"*SINAR*, the mind of the workers"

"*SINAR, penoentoet boeroeh menoejoe socialisme*"
"*SINAR*, prosecutor for workers toward socialism"

March 19, 1924
"*SINAR, memboeka topengnja semoea moesoeh Ra'jat*"
"*SINAR*, unmasking all the enemies of the people"

"*SINAR, adalah tempat pikiran boeroeh*"
"*SINAR*, the mind of the workers"

"*SINAR, melawan pada isapan kapitaal bagi boeroeh*"
"*SINAR*, fighting against the exploitation of capital on workers"

"*Tiap-tiap kaoem boeroeh mesti membatja SINAR!*"
"Every worker must read *SINAR!*"

"*SINAR, masih di daja-oepajaken bagi besar dan baiknja!*"
"*SINAR*, we are still striving for greater and better things!"

"*SINAR, mesti koeat; dari itoe setialah padanja!*"
"*SINAR* must be strong; therefore, be loyal to him!"

"*SINAR, memberi djalan pada Ra'jat oentoek ketinggihan boedi.*"
"*SINAR* gives the people a way to a higher character."

"*Pikirkanlah nasibmoe, hai boeroeh! Sambil mentjari pengetahoean dalam SINAR.*"
"Think of your fate, o workers! While seeking knowledge in *SINAR*."

"*SINAR, soerat kabarnja kaoem boeroeh*"
"*SINAR*, the newspaper of the workers"

"*SINAR menoentoet datangnja persamaan manoesia!*"
"*SINAR* demands equality between humans!"

"*SINAR dengan tidak was-was memberitakan semoea jang bergoena bagi boeroeh.*"
"*SINAR* reports everything useful for workers without hesitation."

"*Sebagai pentingnja makanan bagi toeboeh, begitoelah pentingnja SINAR bagi si Boeroeh.*"
"As important as food is for the body, that is how important *SINAR* is for the workers."

"*Kapak toempoel, asahlah dengan asahan kasar! Otak toempoel, asahlah dengan artikel SINAR!*"
"A blunt axe, sharpen it with a rough sharpener! Dull brain, sharpen it with *SINAR*'s articles!"

"*Sebagai pedoman oentoek pelajar, begitoelah SINAR oentoek Proletar.*"
"As a guide for students, that is *SINAR* for the proletariat."

"*Nasi obatnja peroet lapar, SINAR obatnja fikiran kesasar.*"
"Rice is the medicine for a hungry stomach; *SINAR* is the medicine for a lost mind."

"*Matahari memantjari doenia, SINAR menjinari fikiran kaoem kerdja*"
"The sun brightens the Earth; *SINAR* brightens the workers' mind"

May 9, 1924
"*SINAR tidak soeka moendoer, Sebeloem ia hantjoer.*"
"*SINAR* does not like to retreat before it is destroyed."

"*Keraslah soearanja SINAR, Oentoek memihak si Proletar.*"
"O how loud is the voice of *SINAR*, to defend the proletariat."

"*Angin menioep kekotoran, SINAR membersihkan segenap keboesoekan.*"
"The wind blows away the filth; *SINAR* cleanses all of it."

"*Boeroeng bernjanji bagi peri, SINAR merantasi rantai boemi.*"
"Birds sing for the fairies; *SINAR* will break the chains of the Earth."

"*Fadjar menjingsing terbit matahari, SINAR berdamping memboeka hati.*"
"Dawn breaks clear the rise of the sun; *SINAR* adjoins to open hearts."

"*Tjoeatja terang seloeroeh negeri, SINAR bersihkan kotoran maatschappij*"
"The weather is clear throughout the country; *SINAR* cleans the dirt of the society"

"*Kaboet melajang menarik diri, SINAR datang toendjoek diri*"
"The mist drifts away; *SINAR* comes to show itself"

"*Djanganlah SINAR diroesak dan diboeang-boeang! Berikanlah ia kepada saudara jang ta' mampoe langganan!*"
"Don't let *SINAR* be destroyed and wasted! Give it to a relative who can't afford a subscription!"

"*Tarohlah SINAR, bila soedah dibatja, ditempat-tempat, jang kiranja ia bisa dibatja oleh orang banjak.*"
"Put *SINAR*, when it has been read, in places where it can be read by many people."

SPEAKERS AND CHAIRPERSONS COLLECTED FROM REPORTS ON *OPENBARE VERGADERINGEN* HELD IN 1920–25

There are over 900 names of speakers and chairpersons collected from reports on *openbare vergaderingen* held from 1920 to 1925. The following is the list of female names, Chinese names, and male names. The names include the speakers' places of residence when mentioned in the reports; when it mentions several locations, they are likely several people with the same name.

Female Names

Women who used their husband's names are indicated with "(*woro*)." "*Roro*" refers to unmarried women.

1. Aalijah
2. Andjani
3. Anisa (*roro*)
4. Ati Padmosoemadhio (Malang)
5. Atikah (Bandoeng)
6. Boediprawiro (*woro*)
7. C. Sjamsoe (*woro*)
8. Celestine Chamidin (Poerwodadi)
9. Dj Toegimin (*woro*)
10. Djoeinah
11. Djoeminah (Soekaboemi)
12. Djoewangsih
13. Djojo (*woro*) (Mergotoehoe)

14. Eni (Tjibatoe)
15. Ento (*woro*) (Tjimahi)
16. Innah
17. Isdarwati (Wirosari)
18. Iskandar (*woro*)
19. *Isteri* Rat (Tjiamis)
20. Kadinah
21. Karsi (Tjepoe)
22. Kartadipoera (*woro*) (Tjimahi)
23. Kasim (*woro*)
24. Kasri (*woro*)
25. Kawi (*woro*) (Soematra)
26. Koerlia Djajadiradja
27. Koestini (Kertosono)
28. Kosasih (*woro*)
29. Kotidjah
30. Maimoenah (Lawas)
31. Maimoenah (Soematra)
32. Mardjoen (*woro*)
33. Mardjohan (*woro*) (Semarang)
34. Mardjoko (*woro*)
35. Marijamah (Ambarawa)
36. Marmi
37. Marsini (*roro*)
38. Mintarata (*woro*)
39. Misirah
40. Moedjinah (Soeloer)
41. Moenasiah
42. Moerdiman (*woro*)
43. Moerdinem
44. Moerdini
45. Moersito (*woro*) (Cornelis)
46. Moertini
47. Moesniati
48. Moh. Jasin (*woro*)
49. Moh. Singgih (*woro*)
50. Nafsiah
51. Napsiah
52. Ngadiman (*woro*)

53. Ngaribi (*woro*)
54. Pamintoroto (*woro*)
55. Pranoto (*woro*)
56. Rabiah (Soematra)
57. Rachmat (*woro*)
58. Rakimin (*woro*)
59. Reksokoesoemo (*woro*)
60. Reno (Soematra)
61. Roemelah (*woro*)
62. Roes
63. Roesminah
64. Saarah (Lawas)
65. Sastrodimoelio (*woro*)
66. Sastrosoeprapto (*woro*)
67. Sittie Roesilah
68. Soejatmi (Japara/Mlonggo)
69. Soeketji (Semarang)
70. Soekini
71. Soekirah (Soeloer)
72. Soelijah
73. Soellah
74. Soemarisah
75. Soemarni (Malang)
76. Soeminah
77. Soemodiprodjo (*woro*)
78. Soemoprodjo (*woro*)
79. Soenardi (*woro*) (Semarang)
80. Soepardi (*woro*)
81. Soepijah
82. Soertinah
83. Soetijah
84. Soetiran (Semarang)
85. Soetitah
86. Soewarni
87. Tarjati (Soekaboemi)
88. Tasmijah (*roro*)
89. Tjiptoprawiro (*woro*)
90. Toempoek
91. Wieziettoch-Ardjo

Chinese Names

1. Ong Ek Kiam (Preangan Timoer)
2. Koo Liong Poo
3. Koo Liong Tjiang (Dampit)
4. Koo Tjoen (aka Kho Tjun Wan)
5. Leim Boen Thaij
6. Lie The Ik Bie
7. Liem Boen Thay/Thaij
8. Liem Kiem Tiwie
9. Njo Joe Tik (from Kong Sing Hwee)
10. Oeij Hong Tjoan
11. Oeij Poen Seng (aka Oen Poen Seng; Tjimahi)
12. Oen Poen Seng
13. Tan Ping Tjiat
14. Tan Shee Khing
15. Tan Swie He (Soerabaja)
16. Tan Thaij Tiwan
17. Tjan Sioe Tjoan

Male Names

1. (Mohamad) Djaid
2. A. Raoef (Lawas)
3. A. Thaib (aka A. Thajib)
4. A. Wahab (Soematra)
5. A. Adjis
6. A. Bassach
7. A. Limin (Soematera)
8. A. Moeloeh
9. A. Moeloek
10. A. Moentalib
11. A. Moethalib (Semarang)
12. A. Winanta
13. A. Wiratma
14. Abas
15. Abdoel Moeis
16. Abdoelchalim
17. Abdoerrachman
18. Abdul halim
19. Abdullah

20. Abdullah Fatah
21. Abdulrasid (Semarang)
22. Aboe
23. Aboe Bakar
24. Achmad
25. Achmad Basach
26. Achmad Soebrata
27. Achmad Tajib (Kediri)
28. Achmadbadri
29. Adenan
30. Adikoesoemo (Soerabaja)
31. Adiwidjaja
32. Adiwinata (Tjibatoe)
33. Adiwisastro (Sindanglaoet)
34. Adjraam (Soematera)
35. Adoer (Soekaboemi)
36. Affandi (Manggarai)
37. Agoes Salim (CSI)
38. Ahmad
39. Ahmad (Karangpilang, Surabaya)
40. Ahmadchatib (Soematra)
41. Ahmadsasmita (Tjibatoe)
42. Ahmat (Soerabaja)
43. Alamoeddin (Soematra)
44. Albi
45. Ali (Semarang)
46. Ali Achmad
47. Ali Archam
48. Ali Sastrosoewirjo
49. Alimin
50. Alip Soemarto (Djapara)
51. Aloewi
52. Amat Nawawi (Cornelis)
53. Amat Soeleman
54. Amatbakri
55. Amatsaleh
56. Amattajib
57. Ambijah
58. Ami (Tjimahi)
59. Aminkoesasi (Paree)

60. Amir (Leuwegadjah)

61. Amir (Semarang)

62. Amon

63. Amongsoewarno

64. Anggakoesoema

65. Angsakoesoema

66. Ardij/Ardy Soeroto

67. Arkieman

68. Asmosoenjoto

69. Asnawi

70. Astrolpan

71. Astroredja

72. Ater (Mlonggo)

73. Atjeng (Leuwegadjah)

74. Atmokarjono

75. Atmosardjono

76. Atmosiswojo

77. Atmosoedarmo

78. Atmosoedirdjo

79. Atmosoemarto (Solo)

80. Atmowasito

81. Atmowidjojo

82. Avandie

83. Baars

84. Bachram (Pati)

85. Baharoeddin Said (Padang Pandjang)

86. Baharuddin (Soematra)

87. Bahri (Tjiamis)

88. Bakar S. (Soematra)

89. Basir/Baseer (Djepara)

90. Basiran

91. Basjarroeddin Gafoer (Padang Pandjang)

92. Basoer (Djepara)

93. Bedawi (Pekalongan)

94. Benoe

95. Bergsma (Semarang)

96. Boedihardjo (Cornelis)

97. Boediman

98. Boediprawiro

99. Boedisoetjitro

100. Boedjak (Kalinjamat)
101. Boehari
102. Boekari
103. Boentarman (Indramajoe)
104. Boesro
105. Brahim (Betawi)
106. Bratanata
107. Broto Prajitno
108. Brotoatmodjo
109. Brotoprajitno (Mlonggo)
110. Brotowinotho
111. Ch. M. Pandij
112. Ch. Rachmat
113. Chamidin (Poerwodadi)
114. Chasan Rahmat (Toeren)
115. Dt. M. Besar (Soematra)
116. Dachlan
117. Danadi
118. Danoe (aka Danoewarso, Mlonggo)
119. Danoediwongso
120. Darmono
121. Darmoprawiro
122. Darmosoemarto (Amboeloe, Djember)
123. Darmosoewito (Semarang)
124. Darsono
125. Dartodihardjo
126. Dartowidarmo
127. Dasoeki (Solo)
128. Datoek Mangkoetobesar (Soematra)
129. Dhaham
130. Dhana (Tjibatoe)
131. Dilawilastra
132. Dirdjosoemarto
133. Ditawilastra (Tjiamis)
134. Djadi
135. Djahidi
136. Djajadirdja
137. Djajadiredja
138. Djajakoesman
139. Djajasasmita (Tjimahi)

140. Djajasoemita (Tjimahi)
141. Djajoesman
142. Djamaloeddin (Soematra)
143. Djamaluddin Tamim
144. Djamil
145. Djasmin
146. Djiman
147. Djimat (Randoeblatoeng)
148. Djoedi
149. Djoedohadiwinoto
150. Djoefrie
151. Djoepel
152. Djoeperi
153. Djoepri
154. Djoetri
155. Djoewardi
156. Djojoatmodjo
157. Djojodihardjo
158. Djojodimedjo
159. Djojodirekso
160. Djojodiwirjo (Djoewana)
161. Djojomerkaso
162. Djojopranoto
163. Djojosepoetro (Amboeloe, Djember)
164. Djojosoedarmo
165. Djojosoediro
166. Djojosoewondo
167. Djojotoegimin
168. Djojowijono
169. Djolowijono
170. Doelkamid
171. Dwidjosoemarto
172. E. Karijodimedjo
173. E. Kariodimedjo
174. Endoen (Sindanglaoet)
175. Eni
176. Essom (Betawi)
177. F. Noelik
178. Gandasasmita (Tjimahi)
179. Gazali (Soematera)

180. Gees
181. Goenawan Prawirosardjono (Soerabaja)
182. H. Moh. Imam
183. H. Prajitno (Tjepoe)
184. H. Prawito (Wirosari)
185. H. Acsari
186. H. Azat (Japara)
187. H. Basri (Betawi)
188. H. Raflie
189. H. Sirad
190. H. St Kajo (Soematra)
191. H.A. Djoenaedi
192. H.S. Assor
193. H.S.S. Parpatieh (Soematra)
194. Hadiawiro
195. Hadipranoto/H. Pranoto
196. Hadiprawiro
197. Hadisasmito
198. Hadisoebroto
199. Hadisoemarto
200. Haditomo (Semarang)
201. Hadji Abas
202. Hadji Abdoelkadir
203. Hadji Asaad
204. Hadji Idris
205. Hadji Moehamad Iman (Blitar)
206. Hakan (Soematra)
207. Hamingwardojo
208. Handojomo
209. Handoyomo
210. Harafiah
211. Hardho
212. Hardjaoetomo
213. Hardjodiwongso
214. Hardjooetomo
215. Hardjoprawiro
216. Hardjosoekotjo
217. Hardjosoemarto
218. Hardjosoepardjo
219. Hardjosoeparto (Glenmore)

220. Hardjosoewarjo
221. Hardjosoewarto
222. Hardjosoewignjo
223. Hardjosoewirjo
224. Hardjowijoto
225. Hardjowinoto
226. Hardosoetomo (Salatiga)
227. Haris Soehardja
228. Harjodikromo
229. Harjono
230. Harmili
231. Haroen
232. Hasanbasari
233. Hasanoedin (M. Cornelis)
234. Hatmosoenjoto (Kalinjamat)
235. Hatmowidjojo
236. Heroejono
237. Hoesen
238. Hosen (Bogor)
239. Ibrahim Madjid
240. Idroes (Soematra)
241. Ijam
242. Iksan
243. Imam Prawiro
244. Isdi (Amboeloe, Djember)
245. Isdy Soeroto
246. Iskak
247. Iskander
248. Ismadi
249. Ismail
250. Ismaoen
251. Istohari (Tjiamis)
252. Jatmin
253. Joadohadinoto
254. Joedhohadinoto
255. Joedodinoto (Madioen)
256. Joedohadinoto
257. Joesman
258. Joewono
259. K. Soemardo

260. K. B. Kuslulat
261. Kaboel (Kediri)
262. Kadarisman
263. Kadiroen
264. Kamari
265. Kamidin
266. Kamsir (Betawi)
267. Karawoe
268. Kardjono
269. Kariosantosa
270. Karmawidjaja
271. Karnowidjaja
272. Kartaatmadja
273. Kartadipoera
274. Kartawidjaja
275. Kartawiria (Cimahi)
276. Kartodanoedjo Godek (aka Danoedja; Kalinjamat)
277. Kartodinoedjo
278. Kartodipoero
279. Kartomidjojo
280. Kartono
281. Kartorasad
282. Kartosoedarmo
283. Kartosoewignjo
284. Kartosoewiknjo
285. Kartowinoto
286. Kartowitjitro (aka Moh. Saleh)
287. Karwadi (Mlonggo)
288. Kasanmoestari (aka Moestari)
289. Kasboelah
290. Kasijo
291. Kasman
292. Kasmana
293. Kasmeni
294. Kasmidjan
295. Kasmin (Kalinjamat; Mlonggo; Pati)
296. Kasmo
297. Kasnoer
298. Kastamoen
299. Katimah (Lawas)

300. Kemis (Soekaboemi)

301. Kertosastro

302. Kijahi Abdulchadir (Djepara)

303. Kiswarin

304. Kodir

305. Koendiat

306. Koerdaat

307. Koesmin/Koesnin

308. Koesnan

309. Koesno

310. Koesnomalibari

311. Koesoemowidjo

312. Koesoemowidjojo

313. Koesrin

314. Koewat (Singosari)

315. Koosno

316. Kordaat

317. Krain Baba

318. L. L. Neubessie

319. Lalombo

320. Lando Oemar

321. Lata Ali-Acbar (Betawi)

322. Latjipsastroamidjojo

323. Loekito (Tjepoe)

324. M. Soleh Dj (Padang Pandjang)

325. M. A. A. S. Parpatieh (Soematra)

326. M. B. Amtiram

327. M. B. Mael

328. Machoedoem Sali (Padang Pandjang)

329. Madnaseh (Tjimahi)

330. Maitodirono

331. Makroep

332. Maliki (Tjimahi)

333. Mamesah

334. Mangkoediardjo

335. Mangoen (Bandjaran)

336. Mangoen Sastromaninong (Japara; Mlonggo)

337. Mangoenhardjo

338. Mangoensoehardjo

339. Marah Soetan (Soematra)

340. Marco
341. Mardjoan (aka Mardjohan)
342. Mardjoeki (Salatiga)
343. Mardjoen
344. Mardjoko
345. Marhoem (Padang Pandjang)
346. Marhoen (Soematra)
347. Mariman (Semarang)
348. Markazan/Markasan
349. Markiman
350. Marsait
351. Marsiman
352. Marsino
353. Marsoem (Kramat)
354. Marsoen
355. Martodisoemo
356. Martojo
357. Mashoel
358. Masnoen
359. Mat Rawi
360. Matdosin/Matdasim
361. Matkrani (M. Cornelis)
362. Matoesin (Semarang)
363. Mawardi
364. Mch. Soeparto
365. Md. Praptosoedirdjo
366. Md. Sjamsoe
367. Mhd. Kasan
368. Michrab
369. Mintoroto (Kertosono)
370. Misbach
371. Moch. Ali (Betawi)
372. Moch. Halil
373. Moch. Sarip (SI Tjokro)
374. Mochtadi (aka Sastrohardono)
375. Mochtar (Soekaboemi)
376. Moechtar (Buitenzorg)
377. Moechtar (Soekaboemi; Bogor)
378. Moedjiman/Medjiman
379. Moedjina

380. Moehadi
381. Moehamad Arief (Soematra)
382. Moehammad Tahir
383. Moehtadi (Japara)
384. Moehtadlie Sastrohardjo (Mlonggo)
385. Moeis
386. Moekandar
387. Moekimin
388. Moekmin
389. Moeljowidjojo (Glenmore)
390. Moenaf (Padang)
391. Moeroto Wirjosoemarto
392. Moersito
393. Moes Pardin
394. Moesfa'at
395. Moesirin
396. Moeso
397. Moestakim (Blitar)
398. Moestamir
399. Moestedjo
400. Moestidjo
401. Moetalib
402. Moeto Kalimoen/Moetokalimoen
403. Moh. Bachram
404. Moh. Djaidi
405. Moh. Hasan
406. Moh. Iman
407. Moh. Said
408. Moh. Sanoesi (Bandoeng)
409. Moh. Sidik (Bogor)
410. Moh. Sirat (Tajoe)
411. Moh. Soeebsoetisnasendjaja (Tjibatoe)
412. Moh. Tajib (Paree)
413. Moh. Joesoef
414. Moh. Sadik
415. Mohamad Djaidi
416. Mohamad Djohari
417. Mohamad Jasin
418. Mohamad Tahar
419. M.S. Djaafar (Soematera)

420. Nagli (Bandjaran)
421. Nahootmadja
422. Najoan (Semarang)
423. Nataatmadja (Togogapoe)
424. Nawawi Arief
425. Nawawi Arief (Padang Pandjang)
426. Ngadiman
427. Ngadina
428. Ngadino
429. Ngali (aka Boediprawiro) (Japara; Mlonggo; Bandjaran)
430. Ngaliman (Semarang)
431. Ngalise
432. Ngoesman
433. Nitimarsaid
434. Nitioetomo
435. Noer Ibrahim (Soematra)
436. Noersam
437. Nolowardjojo
438. Notoatmodjo
439. Notohardjo
440. Notosamarata
441. Notosoedarmo
442. Notosoemarto
443. Notosoewandi
444. Notowardojo
445. O. K. Jaman
446. Oding
447. Oele (Leuwegadjah)
448. Oemar (Djokja)
449. Oentoeng
450. Oerip-Asnawi
451. Oesman (Kalinjamat)
452. Oesman Loebis
453. Oesoep Mangoendwiradja
454. Oetojo
455. Oomar (Sindanglaoet)
456. P. Manesah
457. Padma (Togogapoe)
458. Padmodisastro
459. Padmohoetomo

460. Padmosoedijo
461. Padmosoemadjo (Malang)
462. Padmosoemardhjo
463. Padmosoemarto
464. Padmotanojo
465. Paiman
466. Parimin
467. Partaatmadja
468. Partasoeganda
469. Parto Widomo (Malang)
470. Partoatmodjo
471. Partodihardjo
472. Partodiwirjo (Cornelis)
473. Partooetomo
474. Partoredjo
475. Partosimodjo
476. Partosoedarmo
477. Partowi (Malang)
478. Partowidarmo
479. Pasimin (Semarang)
480. Patty (Ambon)
481. Poeradisastra
482. Poerwodihardjo
483. Poerworawiro
484. Poerwowinoto
485. Pontjo (Cornelis)
486. Prajitnowisastro
487. Pramoedihardjo
488. Pramoedjo
489. Pramono
490. Pranotosoedarmo
491. Prapto (Ambarawa; Salatiga)
492. Prawiraadinata
493. Prawiranata
494. Prawirawinata
495. Prawiro (Soekaboemi)
496. Prawiro Soehardjo (Madioen)
497. Prawiro-Adiprodjo
498. Prawiroatmodjo
499. Prawirodihardjo

500. Prawirodiprodjo
501. Prawirohardjo
502. Prawiromedjo
503. Prawiromiastro
504. Prawiroredjo
505. Prawirosantono
506. Prawirosardjono (Soerabaja)
507. Prawirosimoen/Simoen
508. Prawirosoehardjo (Madioen)
509. Prawirowinoto
510. Prawitowerdojo
511. Prawoto
512. Prijodihardjo
513. Prijokoesoemo
514. Projokoesoemo
515. Pronggo Hardjo
516. Rademo
517. Radimin (Tjirebon)
518. Rakimin
519. Ramaja (Soematra)
520. Ramelan
521. Ramlam
522. Rangkajo (Lawas)
523. Ranoewisastro
524. Ratmadji
525. Respati
526. Riedoewan (Mlonggo)
527. Rochmanda
528. Roeslam (Blitar)
529. Roestam (Padang Pandjang)
530. Ronoprawro
531. Roso
532. Rotohadidjojo
533. S. Asanoedin (Meester-Cornelis)
534. S. Atir (Bandjaran)
535. S. Bachrom
536. S. Dasir
537. S. A. Siregar
538. S. H. Pratisto
539. S. H. Rasid/Alirasid

540. Saboer
541. Sadrin (Bogor)
542. Saham
543. Said Alatas (Betawi)
544. Sajoebi
545. Sajoeti
546. Sakariman
547. Salam
548. Saleh Djafar (Padang Pandjang)
549. Saleman (Semarang)
550. Salimin
551. Salimoen
552. Sambik
553. Samiardjo
554. Samin Karto Soewignjo
555. Samirasa
556. Samoedro (Betawi)
557. Samsi
558. Samsoeri (Solo)
559. Sandjojo
560. Sanjoto
561. Sanraip (Tjiamis)
562. Santoso/Santosa
563. Sardjono (Soekaboemi)
564. Sari (Sidoardjo)
565. Sarjoen
566. Sasmito
567. Sastroatmodjo
568. Sastrodarmodjo
569. Sastrodihardjo (Salatiga)
570. Sastrodimoelio
571. Sastrodimoeljo
572. Sastrodirdjo
573. Sastrohardjo (Tjepoe)
574. Sastrokoesoemo
575. Sastromartojo
576. Sastromartono
577. Sastropawiro
578. Sastroprajitno
579. Sastroprawiro

580. Sastroredjo
581. Sastrosoedarmo
582. Sastrosoekarto
583. Sastrosoelias (Keling)
584. Sastrosoemitro
585. Sastrosoeprapto
586. Sastrosoeratmo
587. Sastrosoewirjo (Tjirebon)
588. Sastrowidjojo
589. Sastrowidjono (Solo)
590. Sastrowigoena
591. Satja
592. Sdardjono
593. Sehat
594. Semaoen
595. Semedi
596. S. H. Prawito (Soeloer)
597. Siam
598. Sidik
599. Sidin
600. Singomentolo (aka Sadi)
601. Sirat (Mergotoehoe)
602. Sirroto (aka Sastrodidjojo Kertosono)
603. Sirroto (Kertosono)
604. Siswo
605. Siswojo (Amboeloe, Djember)
606. Slamat
607. Slamat
608. Slamet
609. Sodipoero
610. Sodjojo
611. Soeardjo (Djepara)
612. Soebagijo (propagandist VSTP West Java)
613. Soebakat (Bandoeng)
614. Soebandi
615. Soebeni
616. Soebroto (Semarang)
617. Soedarman
618. Soedarno
619. Soedarso

620. Soedarsono
621. Soedibjo
622. Soediran Hardjowijoto
623. Soedirdjo (Semarang)
624. Soediro (Wirosari; Semarang; Bandoeng)
625. Soedjak
626. Soedjandi
627. Soedjono
628. Soedomo (Indramajoe)
629. Soedono
630. Soedoro
631. Soegeng
632. Soegiman
633. Soegiri
634. Soegono
635. Soehardi (Tjimahi; Solo)
636. Soehirman/Soeherman (Kepandjen)
637. Soejadi (Mlonggo)
638. Soejono
639. Soekadi (Cornelis)
640. Soekadis
641. Soekandar
642. Soekandar
643. Soekantawidjaja
644. Soekardi
645. Soekardono
646. Soekarian
647. Soekarman
648. Soekarno
649. Soekarto
650. Soekartono
651. Soekiban (Randoeblatoeng)
652. Soekiman
653. Soekindar
654. Soekir
655. Soekir (Kertosono)
656. Soekirlan
657. Soekirman
658. Soekirno
659. Soekra (Betawi)

660. Soekrabat
661. Soekrawinata
662. Soekreno
663. Soekria (Togogapoe)
664. Soelaiman (Tjiamis)
665. Soeldjana
666. Soeleman (aka Soeleiman, Malang)
667. Soeljana
668. Soemadi (Djepara; Solo)
669. Soemantri
670. Soemardi
671. Soemardi (Randoeblatoeng)
672. Soemardjo
673. Soemarlan
674. Soemarto (Keling; Djepara)
675. Soemartojo
676. Soemasoebrata
677. Soemedi
678. Soemilto
679. Soemirta (Tjimahi)
680. Soemito
681. Soemitrawinata
682. Soemoatmodjo
683. Soemobasir
684. Soemobroto
685. Soemodiprodjo
686. Soemodisastro
687. Soemoeki
688. Soemoekjat
689. Soemoprawoto
690. Soemoprodjo
691. Soemosoebroto
692. Soemosoedirdjo
693. Soemowijoto (aka Notoprodjo)
694. Soemowikarto
695. Soenardi
696. Soenargo
697. Soenario (Ngandjoek)
698. Soenarjo
699. Soenarno

700. Soenarsi

701. Soenarto (Bandjaran)

702. Soendoro (Semarang)

703. Soengkono

704. Soentil Natasapoetra

705. Soepardan

706. Soepardi

707. Soepardji

708. Soepardjo (Kalinjamat)

709. Soeparjo

710. Soeparman (aka Mangoenboedojo)

711. Soepeno

712. Soepirdi

713. Soepoeman

714. Soepradja

715. Soeprapto (Koewoe)

716. Soeprodjo

717. Soerachman

718. Soeradi

719. Soeradji

720. Soerahman

721. Soerat (Tajoe)

722. Soeratman

723. Soeratmodjo

724. Soerijabrata (Tjimahi)

725. Soerijosepoetro (Betawi)

726. Soeriosepoetro

727. Soerjadi (Tjimahi)

728. Soerjonitihardjo

729. Soerjopranoto

730. Soerjosaprodjo

731. Soeroatmodjo

732. Soerogoetomo

733. Soerosastra

734. Soeroso (Rakitan; Bandoeng)

735. Soeroto

736. Soetadihardjo

737. Soetamiarsa

738. Soetan Djenain

739. Soetarman
740. Soetarno
741. Soetaslekan
742. Soetjipto
743. Soetopawiro
744. Soewandi (Kalibaroe)
745. Soewardjo (Semarang)
746. Soewarno (Betawi; Solo)
747. Soewigjo
748. Soewito
749. Soleh (Semarang)
750. Somomihardjo
751. Somowikarto
752. Sopoero
753. Sosroatmodjo
754. Sosrokardono
755. Sosrosoedarmo
756. Sowowijoto
757. St Batoeah (Soematra)
758. St Radja (Lawas)
759. Sundah
760. T. Mohammad
761. Tadjoewit
762. Taibin (Soematra)
763. Tamjis
764. Tan Malaka
765. Tarhami (Tjiamis)
766. Tarmoedji (Blitar)
767. Tarsimin
768. Tasmian
769. Tasmoen (Blitar)
770. Tirodanoedjo
771. Tirtoatmodjo
772. Tirtomihardjo (Wates)
773. Tirtowardojo
774. Tisna
775. Tjiptoamidjojo
776. Tjiptoprawiro
777. Tjiptorahardjo

778. Tjitrodipoero (Semarang)

779. Tjitrosendjojo

780. Tjitrowidjojo

781. Tjokroaminoto

782. Tjokroatmodjo

783. Tjokroredjo

784. Tjokrosoedoro

785. Tjokrosoemarta

786. Tjokrowardjojo

787. Tjokrowardojo (aka Hadipidjono from Ngandjoek)

788. Tjokrowioto

789. Tobin (Salatiga)

790. Toebin

791. Toha (Mlonggo)

792. Tohir (Semarang)

793. Tolia (Mlonggo)

794. Tomoprawiro

795. W. Samarata

796. W. Sasmito

797. Waliki

798. Walikin

799. Wardojo (Doplang)

800. Wardono

801. Warnomo (Blitar)

802. Warwomo

803. Wasimin

804. Wignjoparmono (Salatiga)

805. Winata

806. Wiradimadja

807. Wiradimardja

808. Wiranta (Leuwegadjah)

809. Wirasoedarmamjardja

810. Wiratmadja

811. Wirjoatmodjo

812. Wirjosoedarmo

813. Wirjosoesastro

814. Wirjosoewondo (Semarang)

815. Wiroastro (Rakitan)

816. Wiroboedojo

817. Wirodarmojo
818. Wirosoeharto (Solo)
819. Wongsoredjo
820. Z. Mohamad (Djepara)

Notes

INTRODUCTION

1. *Woro* (Ms./Mrs.) is a title of respect before a woman's name. Sometimes, it is used before the husband's name to indicate that she is the wife of the said man. *Roro* is used for unmarried women. *Wiro* (Mr.) is used for men.

2. While the conservative count is 1.5 million, Chandra argues that over four million people died in Java during the influenza pandemic of 1918 to 1919. Siddharth Chandra, "Mortality from the Influenza Pandemic of 1918–19 in Indonesia," *Population Studies* 67, no. 2 (2013): 185–93. Adrian Vickers reports that the pandemic led to a series of large-scale strikes occurring in sugar estates in Java and its neighboring islands in 1919 and 1920 involving about 20,000 people. See Adrian Vickers, *A History of Modern Indonesia*, 2nd ed. (Cambridge: Cambridge University Press, 2013). The pandemic in general influenced both the end of World War I and the subsequent protests and labor movements. Laura Spinney, *Pale Rider: The Spanish Flu of 1918 and How It Changed the World* (New York: Public Affairs, 2017).

3. John Downing, *Radical Media: Rebellious Communication and Social Movements* (Thousand Oaks, CA: Sage, 2000), v. Also see Annabelle Sreberny-Mohammadi and Ali Mohammadi, *Small Media, Big Revolution: Communication, Culture, and the Iranian Revolution* (Minneapolis: University of Minnesota Press, 1994); John Downing, "Social Movement Theories and Alternative Media: An Evaluation and Critique," *Communication, Culture & Critique* 1, no. 1 (2008): 40–50; and Bart Cammaerts, Alice Mattoni, and Patrick McCurdy, eds., *Mediation and Protest Movements* (Bristol, UK: Intellect, 2013).

4. Joshua Atkinson, *Alternative Media and Politics of Resistance: A Communication Perspective* (New York: Peter Lang Publishing, 2010).

5. As quoted by Atkinson, *Alternative Media and Politics of Resistance*, 16. Also see Chris Atton, *Alternative Media* (London: Sage, 2002).

6. Raymond Williams, "Communications and Community," in *Resources of Hope: Culture, Democracy, Socialism*, ed. Robin Gable (London: Verso, 1989), 20.

7. Raymond Williams, "Communications and Community," 29–30.

8. This view perceives technology as a social and political entity. See: Langdon Winner, "Do Artifacts Have Politics?," in *The Social Shaping of Technology*, ed. Donald A. MacKenzie and Judy Wajcman (London: Open University Press, 1999 [1980]); Lewis Mumford, *Technics and Civilization* (New York: Harcourt, Brace and Co., 1934). On means of communication as means of production, see Raymond Williams, *Culture and Materialism: Selected Essays* (London: Verso, 2005 [1980]); and Raymond Williams, *Television: Technology and Cultural Form* (New York: Schocken Books, 1975). Marx describes the connection of technology and society as "Technology reveals the active relation of man to nature, the direct process of the production of his life, and thereby it also lays bare the process of the production of the social relations of his life, and of the mental conceptions that flow from those relations." Karl Marx, *Capital Volume I* (London: Penguin Group, 1976 [1867]), 493n4.

9. Chad Kautzer, *Radical Philosophy: An Introduction* (Boulder, CO: Paradigm Publishers, 2015), 21; and Colin Barker, Laurence Cox, John Krinsky, and Alf Gunsvald Nilsen, eds., *Marxism and Social Movements* (Leiden, the Netherlands: Brill, 2013).

10. Nick Dyer-Witheford, *Cyber-Marx: Cycles and Circuits of Struggle in High-Technology Capitalism* (Chicago: University of Illinois Press, 1999), 91–93.

11. See McVey's discussion on her view of newspapers in Ruth T. McVey, *The Rise of Indonesian Communism* (Ithaca, NY: Cornell University Press, 1965), xv–xvi.

12. This book uses the spelling of Djamaloedin Tamin that is consistent with the document released in the period; see "Proces Verbaal, Djamaloedin Tamin," December 13, 1932, from Audrey Kahin's personal collection (also referred to in Audrey Kahin, *Rebellion to Integration: West Sumatra and the Indonesian Polity, 1926-1998* [Amsterdam: Amsterdam University Press, 1999], 293). In the postindependence era, he wrote a memoir under the spelling Djamaluddin Tamim, "Sedjarah PKI Djilid III," 1957. In the secondary literature, his name is spelled "Djamaluddin Tamin"; see Kahin, *Rebellion to Integration*.

13. Cf. Charles Tilly, "Spaces of Contention," *Mobilization: An International Quarterly* 5, no. 2 (2000): 135–59.

14. Rianne Subijanto, "From London to Bali: Raymond Williams and Communication as Transport and Social Networks," *European Journal of Cultural Studies* 27, no. 1 (2024): 129–45, https://doi.org/10.1177/13675494231152886.

15. Nick Couldry, *Media: Why It Matters* (Cambridge, UK: Polity Press, 2020), 13.

16. John Sidel, *Republicanism, Communism, Islam: Cosmopolitan Origins of Revolution in Southeast Asia* (Ithaca, NY: Cornell University Press, 2021).

17. Sebastian Conrad, *What Is Global History?* (Princeton, NJ: Princeton University Press, 2017).

18. McVey, *Rise of Indonesian Communism*, xi.

19. Such as written by J. Th. Petrus Blumberger in one of his trilogy *De communistische beweging in Nederlandsch-Indië* [The communist movement in the Dutch East Indies] (Haarlem, the Netherlands: H. D. Tjeenk Willink & Zoon, 1935).

20. Kahin, *Rebellion to Integration*; Steve Farram, "From 'Timor Koepang' to 'Timor NTT': A Political History of West Timor, 1901–1967" (PhD diss., Charles Darwin University, 2004); Steve Farram, "Revolution, Religion and Magic: The PKI in West Timor, 1924–1966," *Bijdragen tot de Taal-, Land- En Volkenkunde* 158, no. 1 (2002), 21–48; McVey, *Rise of Indonesian Communism*; and Harry A. Poeze, *Tan Malaka: Strijder voor Indonesië's vrijheid, levensloop van 1897 tot 1945* [*Tan Malaka: Fighter for Indonesia's freedom, life history from 1897 to 1945*] ('s-Gravenhage: Martinus Nijhoff, 1976).

21. Takashi Shiraishi, *An Age in Motion: Popular Radicalism in Java, 1912–1926* (Ithaca, NY: Cornell University Press, 1990).

22. Sidel, *Republicanism, Communism, Islam*; and Tim Harper, *Underground Asia: Global Revolutionaries and the Assault on Empire* (Cambridge, MA: Belknap Press, 2021).

23. Klaas Stutje, *Campaigning in Europe for a Free Indonesia: Indonesian Nationalists and the Worldwide Anticolonial Movement 1917–1931* (Copenhagen: NIAS Press, 2019); Kankan Xie, "*Estranged Comrades: Global Networks of Indonesian Communism, 1926–1932*" (PhD diss., University of California Berkeley, 2018); Lin Hongxuan, *Ummah Yet Proletariat: Islam, Marxism, and the Making of the Indonesian Republic* (New York: Oxford University press, 2023); and Oliver Crawford, "Translating and Transliterating Marxism in Indonesia," *Modern Asian Studies* 55, no. 3 (2021): 697–733.

24. Nick Nesbitt, *Universal Emancipation: The Haitian Revolution and the Radical Enlightenment* (Charlottesville: University of Virginia Press, 2008), 1. Emphasis in original.

25. Clifford Siskin and Willliam Warner, eds., *This Is Enlightenment* (Chicago: University of Chicago Press, 2010), 1 and 6–8.

26. Lawrence E. Klein, "Enlightenment as Conversation," in *What's Left of Enlightenment? A Postmodern Question*, ed. Keith Michael Baker and Peter Hanns Reill (Stanford: Stanford University Press, 2001), 154.

27. Works on democratic communication produced in democratic movements and revolutions include: Michael Warner, *The Letters of the Republic: Publication and the Public Sphere in Eighteenth-Century America* (Cambridge, MA: Harvard University Press, 1990); James van Horn Melton, *The Rise of the Public in Enlightenment Europe* (New York: Cambridge University Press, 2001); Max Elbaum, *Revolution in the Air: Sixties Radicals Turn to Lenin, Mao, and Che* (London: Verso, 2002); Michael Denning, *The Cultural Front: The Laboring of American Culture in the Twentieth Century* (London: Verso, 1997); Linda Lumsden, *Black, White, and Red All Over: A Cultural History of the Radical Press in Its Heyday, 1900–1917* (Kent, OH: Kent State University Press, 2014); Frankie Hutton, *The Early Black Press in America, 1827–1860* (Westport, CT: Greenwood Press, 1993); and Todd Vogel, ed., *The Black Press: New Literary and Historical Essays* (New Brunswick, NJ: Rutgers University Press, 2001).

28. A quote by Arthur Walter (1805) in Warner, *Letters of the Republic*, 1.

29. Vivek Chibber, *Postcolonial Theory and the Specter of Capital* (London: Verso, 2013), 53. Also see David Zaret, *Origins of Democratic Culture: Printing Petitions and the Public Sphere in Early-Modern England* (Princeton, NJ: Princeton University Press, 2000).

30. James Van Horn Melton, *The Rise of the Public in Enlightenment Europe* (Cambridge, UK: Cambridge University Press, 2001), 274.

31. Joost Kloek and Wijnand Mijnhardt, *Dutch Culture in a European Perspective*, vol. 2, *1800: Blueprints for a National Community* (New York: Palgrave Macmillan, 2004).

32. René Koekkoek, Anne-Isabelle Richard, and Arthur Weststeijn, "Visions of Dutch Empire: Towards a Long-Term Global Perspective," *Bijdragen en Mededelingen Betreffende de Geschiedenis der Nederlanden* 132, 2 (2017): 88, https://doi.org/10.18352/bmgn-lchr.10342.

33. As quoted in Tim Harper and Sunil Amrith, eds., *Sites of Asian Interaction: Ideas, Networks and Mobility* (Cambridge, UK: Cambridge University Press, 2014), 24.

34. Harper, *Underground Asia*.

35. Rudolf Mrázek, *The Complete Lives of Camp People: Colonialism, Fascism, Concentrated Modernity* (Durham, NC: Duke University Press, 2020).

1. MEANS OF COLONIALISM AS A MEANS OF RESISTANCE

1. C. R. Boxer, *The Dutch Seaborne Empire, 1600–1800* (New York: Knopf, 1965); and Els M. Jacobs and P. Hulsman, *Merchant in Asia: The Trade of the Dutch East India Company during the Eighteenth Century* (Leiden, the Netherlands: CNWS Publications, 2006).

2. Joost Kloek and Wijnand Mijnhardt, *Dutch Culture in a European Perspective*, vol. 2, *1800: Blueprints for a National Community* (New York: Palgrave Macmillan, 2004).

3. Armand Mattelart, *Networking the World, 1794–2000* (Minneapolis: University of Minnesota Press, 1999); Dwayne Roy Winseck and Robert M. Pike, *Communication and Empire: Media Markets and Globalization 1860–1930* (Durham, NC: Duke University Press, 2007).

4. As quoted in Joseph Norbert Frans Marie à Campo, *Engines of Empire: Steamshipping and State Formation in Colonial Indonesia* (Hilversum, the Netherlands: Verloren, 2002), 22.

5. The vital role of transport networks for the Dutch empire is reflected in mainstream literature on Dutch colonialism. À Campo's *Engines of Empire* provides a comprehensive account of how shipping lines promoted the formation of the Dutch colonial state. Likewise, G. Roger Knight's *Sugar, Steam and Steel: The Industrial Project in Colonial Java, 1830–1885* (Adelaide, South Australia: University of Adelaide Press, 2004), highlights the role of transport networks in the economic development of colonial Java. Original documents released by the British military in 1919 titled *Military Report on the*

Netherlands' Possessions in the East Indies, 1919, prepared by the General Staff, War Office, no: L/P&S/20/G71, British Library, London contain thorough information on the types of transports, as well as the routes of transport networks within the Indies archipelago, demonstrating their importance for military and security purposes.

6. On transportation and colonial power, see D. Headrick, "A Double-Edged Sword: Communications and Imperial Control in British India," *Historical Social Research* 35, no. 1 (2010): 51–65; Jody Berland, "Space at the Margins: Colonial Spatiality and Critical Theory after Innis," *TOPIA* 1 (1997): 55–82; Seth Siegelaub, "Preface: A Communication on Communication," in *Communication and Class Struggle*, ed. Armand Mattelart and Seth Siegelaub (New York: International General, 1983), 11–13; and Donal P. McCracken and Ruth E. Teer-Tomaselli, "Communication in Colonial and Post-Colonial Southern Africa," in *Handbook of Communication History*, ed. Peter Simonson, Janice Peck, Robert T. Craig, and John Jackson (New York: Routledge, 2013), 423–36.

7. W. F. Wertheim, *Indonesian Society in Transition: A Study of Social Change* (The Hague: W. van Hoeve Ltd., 1964), 59.

8. Remco Raben, "Epilogue. Colonial Distances: Dutch Intellectual Images of Global Trade and Conquest in the Colonial and Postcolonial Age," in *The Dutch Empire between Ideas and Practice, 1600–2000* ed. René Koekkoek, Anne-Isabelle Richard, and Arthur Weststeijn (Cham: Palgrave Macmillan, 2019), 211.

9. Ann Kumar, "Literary Approaches to Slavery and the Indies Enlightenment: Van Hogendorp's 'Kraspoekol,'" *Indonesia* 43 (1987), 43 and 44.

10. Kumar, "Literary Approaches to Slavery," 47–48, emphasis in the original. Also see Wertheim, *Indonesian Society in Transition*, 239; James Rush, "Journeys to Java: Western Fiction about Indonesia 1600–1980," in *Asia in Western Fiction*, ed. Robin W. Winks and James R. Rush (Honolulu: University of Hawai'i Press, 1990), 138.

11. Kumar, "Literary Approaches to Slavery," 49–51.

12. Kumar, "Literary Approaches to Slavery," 53.

13. Jan Breman, *Mobilizing Labour for the Global Coffee Market: Profits from an Unfree Work Regime in Conial Java* (Amsterdam: Amsterdam University Press, 2015), 103.

14. Dirk van Hogendorp, *Berigt van den Tegenwoordigen Toestand der Bataafsche Bezittingen in Oost-Indiën en den Handel op Dezelve* [Report of the present condition of the Batavian possessions in the East Indies and the trade in them], 2nd ed. (Delft: M. Roelofswaert, 1800), 8 quoted in Breman, *Mobilizing Labour for the Global Coffee Market*, 103.

15. James R. Rush, *Opium to Java: Revenue Farming and Chinese Enterprise in Colonial Indonesia 1860–1910* (Jakarta: Equinox Publishing, 2007 [1990]), 20.

16. Reinier Salverda, "Doing Justice in a Plural Society: A Postcolonial Perspective on Dutch Law and Other Legal Traditions in the Indonesian Archipelago, 1600–Present," *Dutch Crossing* 33, no. 2 (2009): 152–70.

17. Breman, *Mobilizing Labour for the Global Coffee Market*, 178.

18. Breman, *Mobilizing Labour for the Global Coffee Market*, 174.

19. Quoted from John Bastin, *The Native Policies of Sir Stamford Raffles in Java and Sumatra: An Economic Interpretation* (Oxford: Clarendon Press, 1957), 68, in Rush, *Opium to Java*, 20.

20. J. S. Furnivall, *Netherlands India* (Cambridge: Cambridge University Press, 1944), 109–10, quoted in Suzanne Moon, "The Emergence of Technological Development and the Question of Native Identity in the Netherlands East Indies," *Journal of Southeast Asian Studies* 36, no. 2 (2005): 195.

21. Wertheim, *Indonesian Society in Transition*, 93.

22. Pramoedya Ananta Toer, *Jalan Raya Pos, Jalan Daendels* [Postal highway, Daendels Way] (Jakarta: Lentera Dipantara, 2005).

23. C. Fasseur, "The Cultivation System and Its Impact on the Dutch Colonial Economy and the Indigenous Society in Nineteenth-Century Java," in *Two Colonial Empire,* ed. C.A. Bayly and D. H. A. Kolff (Dordrecht, the Netherlands: Martinus Nijhoff Publishers, 1986), 137.

24. Wertheim, *Indonesian Society in Transition.*

25. Colin Divall and George Revill, "Cultures of Transport: Representation, Practice, and Technology," *The Journal of Transport History* 26, no. 1 (2005): 104.

26. On topic of technology and colonial Dutch East Indies, see Antoine Cabaton and Bernard Miall, *Java, Sumatra and the Other Islands of the Dutch East Indies* (London: T. F. Unwin, 1911); Rudolf Mrázek, *Engineers of Happy Land: Technology and Nationalism in a Colony* (Princeton, NJ: Princeton University Press, 2002); Wim Ravesteijn and Jan Kop, *For Profit and Prosperity: The Contribution Made by Dutch Engineers to Public Works in Indonesia, 1800–2000* (Zaltbommel, the Netherlands: Aprilis, 2008); Frances Gouda, "Mimicry and Projection in the Colonial Encounter: The Dutch East Indies/Indonesia As Experimental Laboratory, 1900–1942," *Journal of Colonialism and Colonial History* 1, no. 2 (2000); Suzanne Moon, "Constructing 'Native Development': Technological Change and the Politics of Colonization in the Netherlands East Indies, 1905–1930" (PhD diss., Cornell University, 2000); and Moon, "Emergence of Technological Development."

27. Negeri Sembilan, for example, is largely Minangkabau in terms of the origins of its population.

28. *Official KPM Year Book 1937–1938* (Netherlands Indies: De Unie Batavia-Centrum, n.d.), 5–6.

29. *Official KPM Year Book 1937–1938,* 6.

30. On a study on time and transport in colonial Egypt, see On Barak, *On Time: Technology and Temporality in Modern Egypt* (Berkeley: University of California Press, 2013).

31. See the map in Robert Cribb, *Historical Atlas of Indonesia* (Honolulu: University of Hawai'i Press, 2000), 141.

32. *Official KPM Year Book 1937–1938,* 8–9.

33. On the history of railway, see Joop Oegema and Ado Ladiges, *De stoomtractie op Java en Sumatra* [Steam traction on Java and Sumatra] (Deventer, the Netherlands: Kluwer Technische Boeken, 1982); J. E. Banck, *Geschiedenis der Nederlandsch-Indische spoorweg-maatschappij* [History of the Dutch-Indian railway company] (The Hague: M. J. Visser, 1869).

34. See Cribb, *Historical Atlas,* 140.

35. Wim Ravesteijn, "Between Globalization and Localization: The Case of Dutch Civil Engineering in Indonesia, 1800–1950," *Comparative Technology Transfer and Society* 5, no. 1 (2007): 32–64.

36. Also see Peter Nas and Pratiwo, "Java and de Groote Postweg, la Grande Route, the Great Mail Road, Jalan Raya Pos," *Bijdragen tot de Taal-, Land- en Volkenkunde* 158, no. 4 (2002): 707–25.

37. Toer, *Jalan raya pos,* 24.

38. Toer, *Jalan raya pos,* 22.

39. Toer, *Jalan raya pos,* 21.

40. Toer, *Jalan raya pos.*

41. Toer, *Jalan raya pos.*

42. For specific use of science and technology in advancing sugar industry, see Ulbe Bosma, *Sugar Plantation in India and Indonesia: Industrial Production, 1770–2010* (Cambridge: Cambridge University Press, 2013), 130–63. For general use of science and technology in the Indies, see the sources listed in footnote 26.

43. À Campo, *Engines of Empire,* 361.

44. Ulbe Bosma, "Migration and Colonial Enterprise in Nineteenth Century Java," in *Globalising Migration History the Eurasian Experience (16th-21st Centuries)*, ed. Jan Lucassen and Leo Lucassen (Leiden, the Netherlands: Brill, 2014), 178.

45. Ulbe Bosma, *Sugar Plantation in India and Indonesia: Industrial Production, 1770–2010* (Cambridge: Cambridge University Press, 2013), 107, 130.

46. See map in Cribb, *Historical Atlas*, 54. Also see Ann Laura Stoler, *Capitalism and Confrontation in Sumatra's Plantation Belt, 1870–1979* (New Haven, CT: Yale University Press, 1985).

47. Bosma, *Sugar Plantation in India and Indonesia*, 162.

48. Julinta Hutagalung, "Good Tidings from the Frontiers" (MA thesis, Leiden University, 2008).

49. On *prijaji*, see Ron Hatley, Jim Schiller, Anton Lucas, and Barbara Martin-Schiller, *Other Javas away from the Kraton* (Melbourne: Monash University, 1984).

50. Wertheim, *Indonesian Society in Transition*, 100.

51. See Bosma, *Sugar Plantation in India and Indonesia*; Ahmat Adam, *The Vernacular Press and the Emergence of Modern Indonesian Consciousness (1855–1913)* (Ithaca, NY: Southeast Asia Program, Cornell University, 1995).

52. See William C. Redfield, *The Dutch East Indies: Holland's Colonial Empire* (New York: Guaranty Company, 1922), 24.

53. Vincent Kuitenbrouwer, "The Dutch East Indies during the First World War and the Birth of Colonial Radio," *World History Bulletin* 31, no. 1 (2015): 28.

54. Vincent Kuitenbrouwer, "Propaganda That Dare Not Speak Its Name: International Information Services about the Dutch East Indies, 1919–1934," *Media History* 20, no. 3 (2014): 241.

55. Tim Harper, *Underground Asia: Global Revolutionaries and the Assault on Empire* (Cambridge, MA: Belknap Press, 2021), 522.

56. Kuitenbrouwer, "The Dutch East Indies."

57. Kuitenbrouwer, "The Dutch East Indies," 29.

58. B. van Loon, "Scanning Our Past from Amsterdam: Dr. C. J. de Groot and Long-Distance Radio Telegraphy Between Indonesia and Holland," *Proceedings of the IEEE* 88, no. 11 (2000): 1811.

59. Kuitenbrouwer, "The Dutch East Indies."

60. "Scanning Our Past," 1812.

61. Kuitenbrouwer, "Propaganda That Dare Not Speak Its Name," 244. On the genealogy of telephony in the Dutch East Indies, see Joshua Barker, "Telephony at the Limits of State Control: 'Discourse Networks' in Indonesia" in *Local Cultures and the 'New Asia': The State, Culture, and Capitalism in Southeast Asia*, ed. C. J. W. L. Wee (Singapore: ISEAS, 2002), 158–83.

62. Kuitenbrouwer, "Propaganda That Dare Not Speak Its Name," 247.

63. Kuitenbrouwer, "Propaganda That Dare Not Speak Its Name," 249.

64. Marcia Yonemoto, *Mapping Early Modern Japan: Space, Place, and Culture in the Tokugawa Period, 1603–1868* (Los Angeles: University of California Press, 2003).

65. "Wandkaart van Nederl.-Oost-Indië," accessed February 29, 2016, http://maps.library.leiden.edu/apps/iipview?marklat=0&marklon=0&sid=xz35fe3998397&svid=&code=05336-2&lang=1#focus. Also see Ferjan Ormeling, "School Atlases for a Colonial Society: The Van Gelder/Lekkerkerker School Atlases for the Netherlands East Indies 1880–1952" (paper presented at the International Symposium on "Old Worlds-New Worlds": The History of Colonial Cartography 1750–1950, Utrecht University, Utrecht, the Netherlands, August 21–23, 2006).

66. Kuitenbrouwer, "Dutch East Indies," 30.

67. *Official KPM Year Book 1937–1938*, 50.

68. See "Dempo," Rotterdamsche Lloyd, 1931, object no. F18600, Maritime Museum, Rotterdam, the Netherlands.

69. Bosma, "Migration and Colonial Enterprise," 151.

70. Agustinus Supriyono, "Buruh pelabuhan Semarang: Pemogokan-pemogokan pada zaman kolonial Belanda, revolusi dan republik, 1900–1965" ["Semarang port workers: Strikes during the Dutch colonial era, revolution and republic, 1900–1965"] (PhD diss., Vrije Universiteit Amsterdam, 2008).

71. Migration and cultural contacts have often been approached from a Euro- or Atlantic-centric perspective. A few works seeking to de-Westernize the field include Dirk Hoerder, *Cultures in Contact World Migrations in the Second Millennium* (Durham, NC: Duke University Press, 2011); Patrick Manning and Tiffany Trimmer, *Migration in World History* (New York: Routledge, 2020); and Adam McKeown, "Global Migration, 1846–1940," *Journal of World History* 15, no. 2 (2004): 155–89.

72. Matthias van Rossum, "A 'Moorish World' within the Company: The VOC, Maritime Logistics and Subaltern Networks of Asian Sailors," *Itinerario* 36, no. 3 (2012): 39.

73. Titas Chakraborty, "Slave Trading and Slave Resistance in the Indian Ocean World: The Case of Early Eighteenth-Century Bengal," *Slavery and Abolition* 40, no. 4 (2019): 706–26.

74. Van Rossum, "A 'Moorish World.'" 39, 45. Also see Ravi Ahuja, "Mobility and Containment: The Voyages of South Asian Seamen, c. 1900–1960," *IRSH* 51 (2006): 111–41.

75. Atsushi Ota, "Toward Cities, Seas, and Jungles: Migration in the Malay Archipelago, c. 1750–1850," in *Globalising Migration History: The Eurasian Experience (16th–21st centuries)*, ed. Jan Lucassen and Leo Lucassen (Leiden, the Netherlands: Brill, 2014), 181.

76. Bosma, "Migration and Colonial Enterprise," 153.

77. Bosma, "Migration and Colonial Enterprise," 177. On the list of policies restricting freedom of movement of (suspected) propagandists after the communist revolt, see John Th. P. Blumberger, *De communistische beweging in Nederlandsch-Indië* [The communist movement in the Dutch East Indies] (Haarlem, the Netherlands: H. D. Tjeenk Willink & Zoon, 1935), 151, 160.

78. Bosma notes that over the entire nineteenth century, "145,000 European civilians and 84,000 soldiers, 250,000 indentured laborers for Sumatra . . . [and about] 100,000 free Chinese immigrants and perhaps 10,000 Arabians came to the Dutch East Indies at that time" (Bosma, "Migration and Colonial Enterprise," 152). Also see Ulbe Bosma, *The Making of a Periphery: How Island Southeast Asia Became a Mass Exporter of Labor* (New York: Columbia University Press, 2019).

79. See Anthony Reid, *Southeast Asia in the Age of Commerce 1450–1680* (New Haven, CT: Yale University Press, 1993 [1988]); and R. E. Elson, *The End of the Peasantry in Southeast Asia: A Social and Economic History of Peasant Livelihood, 1800–1990s* (London: Macmillan Press Ltd., 1997).

80. In the late 1950s, Sarbupri, a trade union of plantation estate workers, was the largest trade union in the country and was linked to the PKI.

81. Onghokham, "Chinese Capitalism in Dutch Java," *Southeast Asian Studies* 27 (1989): 162.

82. M. R. Fernando, "Dynamics of Peasant Economy in Java at Local Levels," in *Nineteenth and Twentieth Century Indonesia: Essays in Honour of Professor J.D. Legge*, ed. David P. Chandler and M. C. Ricklefs (Clayton, Australia: Monash University, 1986), 97–121. Also see Wertheim, *Indonesian Society in Transition*, 98.

83. Rush, *Opium to Java*, 23.

84. Onghokham, "Chinese Capitalism," 162.

85. Wertheim, *Indonesian Society in Transition*, 254.

86. Approximately 146,696,000 passengers traveled by railroad that year: 129,736,000 in Java and Madura and 16,960,000 in the Outer Islands. Of the total Java passengers, 0.1 percent were first class passengers; 2.5 percent were second class; 21.16 percent were third class; and 76.16 percent were fourth class, with the third and fourth class primarily used by the indigenous population. These were for passengers using services from the following: *Staatsspoorwegen, Nederlandsch-Indische spoorweg maatschappij*, and Urban trams in Batavia. Gerrit J. Knaap, *Changing Economy in Indonesia: A Selection of Statistical Source Material from the Early 19th Century up to 1940*, ed. P. Boomgaard, vol. 9, *Transport 1819–1940* (Amsterdam: Royal Tropical Institute [KIT], 1991), 30–31.

87. The table "Indigenous Population by Residency, 1912–1942" in P. Boomgaard and A. J. Gooszen, *Changing Economy in Indonesia: A Selection of Statistical Source Material from the Early 19th Century up to 1940*, ed. P. Boomgaard, vol. 11, *Population Trends 1795–1945* (Amsterdam: Royal Tropical Institute [KIT], 1991), 120, did not include the number for 1929. So, I predicted this number by calculating the average growth of the population between 1921 and 1928, which was 366,176 people, and then added this number to the number of the population in 1928, resulting in 37,027,564.

88. Multatuli, *Max Havelaar or the Coffee Sales of the Netherlands Trading Company*, translated by W. Siebenhaar (New York: Alfred A. Knopf, 1927 [1860]), 44. Italics in original.

89. Cribb, *Historical Atlas*, 122.

90. Sartono Kartodirdjo, *Protest Movements in Rural Java: A Study of Agrarian Unrest in the Nineteenth and early Twentieth Centuries* (London: Oxford University Press, 1973).

91. "Betreffende de rechtspersoonlijkheid van de vereeniging 'Sarikat Postel'" [Concerning the legal personality of the association "Sarikat Postel"], no. 2082, Inventaris Arsip Post-, Telegraaf- en Telefoondienst 1817–1950 (henceforth PTT), Arsip Nasional Republik Indonesia (henceforth ANRI), Jakarta, Indonesia; and "Verwijdering van communisten bij den Sarekat Postel dienst uit den dienst" ["Removal of communists from the Postal Union service"], no. 2086, PTT, ANRI, Jakarta.

92. "Aangestelde afscrift schrijven 'Propaganda vergadering van het inlandse Personeel van de P.T.T te Soekaboemi'" [Appointed to write copy "Propaganda meeting of the native staff of the P.T.T in Soekaboemi"], no. 2521, PTT, ANRI, Jakarta.

93. "Toezicht op communistische landsdienaren over Onderwijs en Opleidingvergadering 'Sarikat Rajat' en Sarikat Postel" [Supervision of communist national servants on education and training meeting 'Sarikat Rajat' and Sarikat Postel], no. 2089, PTT, ANRI, Jakarta.

94. "Toelichting op het art 'Ckt Soerabaja djadi Dictatuur S.P.', voorkomende in Soeara Postel no. 5/6" [Explanation of the article "Ckt Soerabaja became S. P. dictatorship," appearing in Soeara Postel no. 5/6], no. 2095, PTT, ANRI.

95. "Toezending artikel uit Soeara Kita" ["Article sent from Soeara Kita"], no. 2529, PTT, ANRI, Jakarta.

96. "Toezicht op communistische landsdienaren."

97. "Beslaglegging op copie Soeara Postel te Solo" [Seizure of copy of Soeara Postel in Solo], no. 2094, PTT, ANRI, Jakarta.

98. "Toelichting op het art 'Ckt Soerabaja djadi Dictatuur S.P.'"

99. For a brief history of VSTP, see Hoofdbestuur, *Poesaka V.S.T.P.* [The heritage of V.S.T.P] (Semarang, Indonesia: Typ. Drukkerij V.S.T.P., April 1923), in Collectie Documentatiebureau voor Overzees Recht ca 1900–1958, 144, inv. no. 2.20.61, Nationaal Archief, the Hague. Semaoen began his leadership in VSTP at the age of sixteen when he quit his job as a railway worker to work full time as a trade union activist.

100. Hoofdbestuur, *Poesaka V.S.T.P.*, 144.

101. Supriyono, *Buruh Pelabuhan Semarang*, 107; John Ingleson, *In Search of Justice: Workers and Unions in Colonial Java, 1908–1926* (Singapore: Oxford University Press, 1987), 78.

102. Hoofdbestuur, *Poesaka V.S.T.P.*, 144.

2. RED ENLIGHTENMENT IN THE ROARIN' TWENTIES

1. Kenji Tsuchiya, "Javanology and the Age of Ranggawarsita: An Introduction to Nineteenth-Century Javanese Culture," in *Reading Southeast Asia*, ed. Takashi Shiraishi (Ithaca, NY: Cornell University Press, 1990), 75–108; and Theodore G. Th. Pigeaud, *Literature of Java: Catalogue Raisonné of Javanese Manuscripts* (Leiden, the Netherlands: University of Leiden Library, 1967).

2. Tsuchiya, "Javanology and the Age of Ranggawarsita," 81, 106.

3. Benedict Anderson, *Imagined Communities: Reflections on the Origin and Spread of Nationalism* (London: Verso, 1983), 137, quoted in Kenji Tsuchiya, "Kartini's Image of Java's Landscape," *East Asian Cultural Studies* 25 (1986): 64.

4. Tsuchiya, "Kartini's Image," 66–69.

5. Didi Kwartanada, "*Bangsawan Prampoewan*: Enlightened Peranakan Chinese Women from Early Twentieth Century Java," *Wacana* 18, no. 2 (2017): 422–54.

6. Peter Keppy, "Southeast Asia in the Age of Jazz: Locating Popular Culture in the Colonial Philippines and Indonesia," *Journal of Southeast Asian Studies* 44, no. 3 (2013): 444–64; and Bart Barendregt, Peter Keppy, and Henk Schulte Nordholt, *Popular Music in Southeast Asia: Banal Beats, Muted Histories* (Amsterdam: Amsterdam University Press, 2017).

7. Barendregt et al., *Popular Music in Southeast Asia*, 28; Keppy, "Southeast Asia in the Age of Jazz," 458; and Philip Bradford Yampolsky, "Music and Media in the Dutch East Indies: Gramophone Records and Radio in the Late Colonial Era, 1903–1942" (PhD diss., University of Washington, 2013), 251.

8. Yampolsky, "Music and Media in the Dutch East Indies," 383; and Barendregt et al., *Popular Music in Southeast Asia*, 24–25.

9. Dafna Ruppin, *The Komedi Bioscoop: The Emergence of Movie-Going in Colonial Indonesia, 1896–1914* (Bloomington: Indiana University Press, 2016).

10. Tjempaka-Pasoeroean, "Aliran djaman atau seorang gadis yang sengsara" [The flow of time or a girl who suffers], *Api*, August 6–September 10, 1925.

11. Tjempaka-Pasoeroean, "Aliran djaman," chapter 14.

12. H. M. J. Maier, "Introduction Written in the Prison's Light: *The Story of Kadirun*," in Semaoen, *The Story of Kadirun*, trans. Ian Campbell and John H. McGlynn (Jakarta: Lontar Foundation, 2014), vii.

13. Benedict Anderson, *Language and Power: Exploring Political Cultures in Indonesia* (Ithaca, NY: Cornell University Press, 1990); James T. Siegel, *Fetish, Recognition, Revolution* (Princeton, NJ: Princeton University Press, 1997); Tsuchiya, "Kartini's Image"; and Ruth McVey, "Building Behemoth: Indonesian Constructions of the Nation-State," in *Making Indonesia*, ed. Daniel S. Lev and Ruth McVey (Ithaca, NY: Southeast Asia Program Publications, Cornell University, 1996), 12.

14. A. van den Bergh, "Multatuli and Romantic Indecision," *Canadian Journal of Netherlandic Studies/Revue Canadienne d'Etudes Néerlandaises* 13, no. 2 (1992): 40.

15. Carl Niekerk, "Rethinking a Problematic Constellation: Postcolonialism and Its Germanic Contexts (Pramoedya Ananta Toer/Multatuli)," *Comparative Studies of South Asia, Africa and the Middle East* 23, nos. 1 & 2 (2003): 63.

16. Anne-Marie Feenberg, "'Max Havelaar': An Anti-Imperialist Novel," *MLN* 112, no. 5 (1997): 817–35.

17. Benedict Anderson, "*Max Havelaar* (Multatuli, 1860)," in *The Novel*, ed. Franco Moretti, vol. 2, *Forms and Themes* (Princeton, NJ: Princeton University Press, 2006), 459. For Multatuli's influence on Sigmund Freud's ideas on the sexual enlightenment of children, see Carl Niekerk, "Race and Gender in Multatuli's *Max Havelaar* and *Love Letters*," in *One Hundred Years of Masochism: Literary Texts, Social and Cultural Contexts*, ed. Carl Niekerk and Michael C. Finke (Amsterdam: Brill, 2000), 171.

18. Danilyn Rutherford, "Unpacking a National Heroine: Two Kartinis and Their People," in *Appropriating Kartini: Colonial, National and Transnational Memories of an Indonesian Icon*, ed. Paul Bijl and Grace V. S. Chin (Singapore: ISEAS Publishing, 2020), 107.

19. Rutherford, "Unpacking a National Heroine," 112.

20. Publications of Kartini consulted: Raden Adjeng Kartini, *Door duisternis tot licht* [From darkness to light] ('S-Gravenhage: N. V. Electrische Drukkerij "Luctor et Emergo," 1912), https://www.gutenberg.org/cache/epub/35220/pg35220-images.html; Raden Adjeng Kartini, *Letters of a Javanese Princess*, trans. Agnes Louise Symmers (London: Duckworth, 1921); and Raden Adjeng Kartini, *Habis gelap terbitlah terang*, trans. Empat Saudara (Weltevreden, Indonesia: Balai Poestaka, 1922). The Malay translators, Empat Saudara (Four Brothers), are likely four Indonesian scholars from Pariaman, West Sumatra, who earned education in the Netherlands: Baginda Djamaloedin Rasad, Baginda Zainoedin Rasad, Baginda Dahlan Abdoellah, and Soetan Moehammad Zain; see Suryadi, "Asal-Usul 'Habis gelap terbitlah terang,'" *sumbarsatu.com*, April 21, 2022, https://sumbarsatu.com/berita/27821-asalusul-habis-gelap-terbitlah-terang. I am thankful for Herdiana Hakim for sharing the 1922 Malay copy.

21. Tsuchiya, "Javanology and the Age of Ranggawarsita"; Benedict Anderson, *Language and Power*, 241–70.

22. Anderson, *Language and Power*, 241–45.

23. Michael Hawkins, "Exploring Colonial Boundaries: An Examination of the Kartini-Zeehandelaar Correspondence," *Asia-Pacific Social Science Review* 7, no. 1 (2007): 1.

24. Kartini, *Letters of a Javanese Princess*, 38.

25. Anderson, *Language and Power*, 244.

26. John Nery, *Revolutionary Spirit: Jose Rizal in Southeast Asia* (Singapore: Institute of Southeast Asian Studies, 2011), xii.

27. Reynaldo Ileto, *Pasyon and Revolution: Popular Movements in the Philippines, 1840–1910* (Manila: Ateneo de Manila University Press, 1979), 200. For further reading on enlightenment and the Philippines, see: David Wurfel, *Filipino Politics: Development and Decay* (Quezon City: Ateneo de Manila University Press, 1991), 8, 44; Jose S. Arcilla, "The Enlightenment and the Philippine Revolution," *Philippine Studies* 39, no. 3 (1991), 358–73; Cesar Majul, *The Political and Constitutional Ideas of the Philippine Revolution* (Quezon City: University of the Philippines Press, 1967); and John N. Schumacher, *The Propaganda Movement, 1880–1895: The Creation of a Filipino Consciousness, The Making of the Revolution* (Quezon City: Ateneo de Manila University Press, 1997).

28. Pankaj Mishra, *From the Ruins of Empire: The Intellectuals Who Remade Asia* (New York: Farrar, Straus and Giroux, 2012), 1–9.

29. Mishra, *From the Ruins of Empire*, 2.

30. Toer, *Sang Pemula*, 120–131.

31. Later, these enlightenment and French Revolution ideals influenced Pramoedya Ananta Toer's seminal work, notably *Bumi manusia* (Jakarta: Hasta Mitra, 1980). In a way, the loop continues from Dekker to Kartini to Tirto to Pram. See Niekerk, "Rethinking a Problematic Constellation," 62–63.

32. Anderson, *Language and Power*, 153.

33. See also the definition in Takashi Shiraishi, *An Age in Motion: Popular Radicalism in Java, 1912–1926* (Ithaca, NY: Cornell University Press, 1990), xx.

34. Onghokham, "The Inscrutable and the Paranoid: An Investigation into the Sources of the Brotodiningrat Affair," in *Southeast Asian Transitions: Approaches through Social History*, ed. Ruth Thomas McVey, Adrienne Suddard, and Harry Jindrich Benda (London: Yale University Press, 1978): 117.

35. Ahmat Adam, *The Vernacular Press and the Emergence of Modern Indonesian Consciousness (1855–1913)* (Ithaca, NY: Southeast Asia Program, Cornell University, 1995), 176.

36. *Sama rasa* (same-feeling) and *sama rata* (same-equal) were important vocabularies of the movement, popularized first by Marco Kartodikromo in his article "Sama rata sama rasa" published in *Pantjaran warta* on April 14, 1917. They were Javanese expressions that have a similar meaning to the communist concept of "equality" in terms of both the redistribution of material resources and the recognition of culture and identity. Ahmat Adam, "Mas Marco Kartodikromo dalam perjuangan 'Sama rata sama rasa,'" *Jurnal Kinabalu* 3 (1997): 10.

37. Djoeinah, "Geraknja S.I. perampoean semarang" [The movement of the women of SI Semarang], *Sinar Hindia,* December 20, 1920.

38. Djoeinah, "Zaman ini" [This age], *Sinar Hindia,* December 7, 1920.

39. "Verslag vergadering S.I. Woro Oengaran" [Report meeting SI *woro* Oengaran], *Sinar Hindia,* October 11, 1920.

40. Djoeinah, "Beladjar ta' mengasingkan diri!" [Learning to not isolate oneself], *Sinar Hindia,* February 7, 1921.

41. Djoeinah, "Rintangan" [Challenges], *Sinar Hindia,* July 4, 1921.

42. Karl Marx and Friedrich Engels, *Manifest kommunist oleh Karl Marx dan Friedrich Engels* [Communist manifesto by Karl Marx and Friedrich Engels], translated and introduced by Partondo (Semarang, Indonesia: Type Drukkerij VSTP, 1923), access no. 2.20.61, no. 147, Inventaris van de Collectie Documentatiebureau voor Overzees Recht, 1894–1963, Nationaal Archief, the Hague.

43. After these two versions, other translations of *Communist Manifesto* would be published multiple times beginning at the height of the national revolution in 1946 until 1964. Thanks to Oliver Crawford for collecting and sharing this complete list: *Manifest kommunist,* trans. Partondo (Semarang, Indonesia: VSTP Press, 1923); *Manifest kommunist,* trans. Axan Zain [Subakat] (Semarang, Indonesia: VSTP Press, 1925); *Manifest kommunist,* trans. A. Z. Dahlamy (Padang Pandjang, Indonesia: Badezst, 1946); *Manifes komoenis,* trans. Saudara X (Jogjakarta, Indonesia: Poestaka Proletar, 1946); and *Manifes Partai Komunis,* trans. D. N. Aidit, M. H. Lukman, A. Havil, P. Pardede, and Njoto (Jakarta: Jajasan Pembaruan, first edition 1948, second edition 1952, third edition 1959, fourth edition 1960, fifth edition 1964). The 1948 edition was done to coincide with the centennial of the original publication of Karl Marx's *The Manifesto of the Communist Party* in 1848. The team was led by Rollah Sjarifah, M. H. Lukman's sister, who was assisted by Aidit, Lukman, and Njoto. Soerjono and Benedict Anderson, "On Musso's Return," *Indonesia,* no. 29 (1980), 73. It is peculiar that Rollah's name was not mentioned on the book as one of the translators. Rollah would continue playing a role in the translation of other communists' work, including by Shaoqi Liu, Vladimir Lenin, Joseph Stalin, and Klement Gottwald. Shaoqi Liu, *Tentang Watak Klas Manusia* [On the human's class character], trans. Rollah Sjarifah (Jakarta: Jajasan Pembaruan, 1956); Vladimir Ill'ich Lenin, *Tjiri-Tjiri Tertentu Perkembangan Bersedjarah daripada Marxisme* [The particular characteristics of the historical development of Marxism], trans. Rollah Sjarifah (Jakarta: Jajasan Pembaruan, 1957); Lenin, *Dua Taktik Sosial-Demokrasi di dalam Revolusi Demokratis,* [Two Social-Democratic tactics in a democratic revolution] trans. Rollah Sjarifah (Jakarta: Jajasan Pembaruan, 1957); Lenin, *Tentang Adjaran-Adjaran Karl Marx* [On Karl Marx's teachings], trans. D. N. Aidit and Rollah Sjarifah (Jakarta: Jajasan Pembaruan, 1955); Losif Vissarionovic Stalin, *Tentang*

tugas wakil rakjat dan kaum pemilihnya [On the duties of the people's representatives and their voters], trans. Rollah Sjarifah (Jakarta: Pembaruan, 1954); Vladimir Ill'ich Lenin, *Kepada kaum miskin desa: Suatu pendjelasan bagi petani-petani tentang jang Dikehendaki Kaum Sosial-Demokraat* [To the village poor: An explanation for farmers of what the Social Democrats want], trans. Rollah Sjarifah (Jakarta: Jajasan Pembaruan, 1958); and Klement Gottwald, *Tentang musuh-musuh negara, partai dan rakjat* [Concerning the enemies of the state, party and people], trans. Rollah Sjarifah (Djakarta: Pembaruan, 1955). While the Indonesian communists seem to translate numerous communist-related works, especially during the period of postindependence Indonesia, the translation of Marx's works was meager by comparison. Outside of the *Communist Manifesto* and a chapter or two from Marx's *Capital*, seemingly no other works by him were translated by the Indonesian communist movement both in the early period of 1920s and the period after independence before the New Order era. One report by Semaoen in 1927 from Russia states that two translation projects, Lenin's "State and Revolution" and Stalin's book on Leninism, were underway but it is not clear if they were ever completed. "Minutes of the Anglo-American Secretariat Meeting of December 29, 1927," 6, collection ID: ARCH01744, no. 4, Archief Komintern—Partai Komunis Indonesia, International Institute of Social History (henceforth IISH), Amsterdam.

44. Soeradi, "Apakah Bolsjewisme itoe?" [What is bolshevism?], *Sinar Hindia*, February 19, 1920; "Revolusioner atau evolusioner" [Revolutionary or evolutionary], *Api*, September 17, 1925; and Karl Radek, "Socialisme: Dari angan-angan hingga djadi pengetahoean (wetenschap)" [Socialism: From fantasy to knowledge (science)], *Api*, November 22, 1925.

45. This special Sunday edition was new; however, it did not appear in the subsequent months until the closing of *Api* in April 1926.

46. Oliver Crawford, "Translating and Transliterating Marxism in Indonesia," *Modern Asian Studies* 55, no. 3 (2021): 697–733.

47. James T. Siegel, *Fetish, Recognition, Revolution* (Princeton, NJ: Princeton University Press, 1997), 19.

48. "Si manis kontra si merah: Kapital goela setengah mati" [The sweet versus the red: The capitalist sugar is dying], *Api*, January 29, 1926.

49. Adrian Vickers, *Balinese Art: Paintings and Drawings of Bali, 1800–2010* (North Clarendon, VT: Tuttle Publishing, 2012), 38.

50. S. Tjitrosoebono, "Soekakah bersatoe hati?" [Do we want to reconcile?], *Sinar Hindia*, July 26, 1921. Italics in the original. Non-Malay words are kept in original in this translation to show the mix of language.

51. Adam, *Vernacular Press*, 9–10. Low Malay was to be differentiated from High Malay, which was a more proper and pure form of Malay that the Dutch government promoted to counter Low Malay, also known as "Melayu pasar" (Market Malay).

52. "Perkoempoelan soerat kabar dan Perskantoor" [The association of newspapers and persbureau], *Api*, November 21, 1925.

53. "Kekuasaan surat kabar" (The power of newspapers), *Sinar Hindia*, April 5, 1920. This was the highest position in the Dutch colonial government. He was responsible to report directly to the Dutch Cabinet.

54. See a complete list of songs, slogans, and cartoons appearing in *SH/Api* from 1918 to 1926 in the appendices.

55. Literature on the political significance of slogans: Joel D. Aberbach and Jack L. Walker, "The Meanings of Black Power: A Comparison of White and Black Interpretations of a Political Slogan," *American Political Science Review* 64, no. 2 (1970): 367–88; Anna L. Ahlers and Gunter Schubert, "'Building a New Socialist Countryside'—Only a Political Slogan?," *Journal of Current Chinese Affairs* 38, no. 4 (2009): 35–62; and Ludwig

Bieler, "A Political Slogan in Ancient Athens," *The American Journal of Philology* 72, no. 2 (1951): 181–84.

56. "Report of Comrade Semaoen to British Secretariat Meeting of March 8, 1927 on Indonesian Question," 2–3, collection ID: ARCH01744, no. 16, Archief Komintern—Partai Komunis Indonesia, IISH, Amsterdam.

57. "Minutes of Meeting of British Secretariat Held June 23rd, 1927," 1, collection ID: ARCH01744, no. 16, Archief Komintern—Partai Komunis Indonesia, IISH, Amsterdam.

58. "Replies on the Twelve Questions," 1 and 6, collection ID: ARCH01744, no. 16, Archief Komintern—Partai Komunis Indonesia, IISH, Amsterdam.

59. "Indonesian Conference, July 22, 1923," 6, collection ID: ARCH01744, no. 16, Archief Komintern—Partai Komunis Indonesia, IISH, Amsterdam.

60. P. J. Brummett, *Image and Imperialism in the Ottoman Revolutionary Press: 1908–1911* (Albany: SUNY Press, 2000).

61. "Report of Comrade Darsana to India Sub-Secretariat," May 6, 1926, collection ID: ARCH01744, no. 16, Archief Komintern—Partai Komunis Indonesia, IISH, Amsterdam.

62. Marco, "Nasib Kami" [Our fate], *Sinar Hindia*, July 11, 1921.

63. "Tjamboek" is a popular pen name during this period, and it is unclear who wrote the contributions to *SH/Api* under that name. The whip symbol might be used to fight against the most notorious racist and reactionary newspaper of the Dutch, *De zweep* (The Whip). Thanks to Benedict Anderson for pointing this out in our email exchange. "Si Tjamboek" or "De Zweep" is also the penname of a skilled journalist of Italian descent, Dominique Willem Beretty, who later founded the news agency Aneta. He was ten years younger than Tirtoadhisoerjo but was involved in terrorizing him in the Brunsveld van Hulten scandal circa 1909. Pramoedya Ananta Toer, *Sang Pemula* (Jakarta: Hasta Mitra, 1985). It is possible that Beretty himself contributed his writings to *Sinar Hindia* or other writers inspired by him borrowed the same alias. Peter Keppy also notes that Kwee Thiam Thjing, a young *peranakan* Chinese journalist in Surabaya who had a great sympathy for Indonesian nationalism, used the penname "Tjamboek Berdoeri" (Spiked Whip). Keppy, "Southeast Asia in the Age of Jazz," 460.

64. Thanks to Jan van der Putten for pointing this out.

65. John A. Lent, "Cartooning in Indonesia: An Overview," in *Southeast Asian Cartoon Art: History, Trends and Problems*, ed. John A. Lent (Jefferson, NC: McFarland, 2014), 10.

66. Anderson, *Language and Power*, 162.

67. Lent, "Cartooning in Indonesia," 9.

68. Lent, "Cartooning in Indonesia"; Anderson, *Language and Power*; and Muliyadi Mahamood, "The Development of Malay Editorial Cartoons," *Southeast Asian Journal of Social Science* 25, no. 1 (1997): 37.

69. "Kapitalisme atau kommunisme-kah jang membikin kemelaratan" [Capitalism or communism that leads to sufferance], *Sinar Hindia*, January 12, 1924.

70. Mahamood, "Development of Malay Editorial Cartoons," 53–54.

71. A report to Comintern quoting the *Handbook voor cultuur and handels onderne-mingen* (issued in 1927) shows that most capital belonged to European companies from Holland, England, Belgium, Switzerland, France, and Germany with Holland holding the most at 60 percent. Outside of Europe, the United States held 2 percent, China 10 percent, Japan 2 percent, and native Indonesians 0.8 percent, affirming the common belief at the time that most capitalists in the Indies were of foreign background. "The Foreign Capitals of Indonesia," 4, collection ID: ARCH01744, no. 11, Archief Komintern—Partai Komunis Indonesia, IISH, Amsterdam.

72. This is a saying by Sudjojono, an Indonesian critic of "beautiful Indies" landscapes who established a new school of realism, quoted in Tsuchiya, "Kartini's Image," 68.

73. Mahamood, "The Development of Malay Editorial Cartoons," 40.

74. "Crocodiles Created a *Vergadering*," *Sinar Hindia*, June 6, 1924.

75. Anderson, *Language and Power*, 156.

76. Peter Boomgaard, *Frontiers of Fear: Tigers and People in the Malay World, 1600–1950* (New Haven, CT: Yale University Press, 2001), 6.

77. Boomgaard, *Frontiers of Fear*, 6.

78. Boomgaard, *Frontiers of Fear*, 61. Although rare, positive view on tigers as a "friend" could also be found that does not see them as mere deadly enemies. Boomgaard, *Frontiers of Fear*, 59.

79. Boomgaard, *Frontiers of Fear*, 108, 136.

80. Boomgaard, *Frontiers of Fear*, 161. Also see Robert Wessing, "A Tiger in the Heart: The Javanese Rampok Macan," *Bijdragen tot de Taal-, Land- en Volkenkunde* 148 (1992), 289–291.

81. Boomgaard, *Frontiers of Fear*, 162.

82. Tsuchiya, "Kartini's Image," 68.

83. Anderson, *Language and Power*, 169.

84. Vickers, *Balinese Art*, 42–43.

85. For further analysis on Sin Po, see Leo Suryadinata, *Peranakan Chinese Politics in Java, 1917–1942* (Singapore: Singapore University Press, 1981); and Peter Post, Juliette Koning, and Marleen Dieleman, eds., *Chinese Indonesians and Regime Change* (Leiden, the Netherlands: Brill, 2011).

86. Thanks to Jan van der Putten and Meghan Forbes for a discussion on the materiality of these cartoons.

87. Claudine Salmon, *Sastra Indonesia Awal: Kontribusi Orang Tionghoa* (Jakarta: Kepustakaan Populer Gramedia, 2010), 15–16. See also Claudine Salmon, *Literature in Malay by the Chinese of Indonesia: A Provisional Annotated Bibliography* (Paris: Editions de la Maison des sciences de l'homme, 1981).

88. Razif, *Bacaan liar budaya dan politik pada zaman pergerakan* [Wild reading of culture and politics in the era of the movement] (Jakarta: Edi Cahyono Experience, 2005), 2. Also see Hilmar Farid and Razif, "Batjaan Liar in the Dutch East Indies: A Colonial Antipode," *Postcolonial Studies* 11, no. 3 (2008): 277–92; and A. Teeuw, *Modern Indonesian Literature* ('s-Gravenhage: Martinus Nijhoff, 1967), 15–17.

89. Razif, *Bacaan Liar*, 2.

90. Semaoen, *Hikajat Kadiroen* [The tale of Kadiroen] (Semarang, Indonesia: Kantoor P.K.I, 1920), housed in the Kroch Library, Cornell University, Ithaca, NY. A recent edition is Semaoen, *Hikayat Kadiroen: Sebuah novel* [The tale of Kadiroen: A novel] (Yogyakarta, Indonesia: Yayasan Bentang Budaya, 2000).

91. Soemantri, *Rasa merdika: Hikajat Soedjanmo* [A sense of independence: The tale of Soedjanmo] (Semarang, Indonesia: VSTP, 1924), Kroch Library, Cornell University, Ithaca, NY.

92. Tjempaka-Pasoeroean, "Aliran djaman atau seorang gadis yang sengsara" [The flow of time or a girl who suffers], *Api*, September 1, 1925.

93. "Dilepas karena tertoedoeh lid S.R" [Released after accused of being an SR member], *Api*, July 23, 1925.

94. Maier, "Introduction Written in the Prison's Light," ix.

95. Maier, "Introduction Written in the Prison's Light," ix, xxii.

96. See Maier's discussion on didacticism in Maier, "Introduction Written in the Prison's Light," xxvi.

97. A Teeuw, "The Impact of Balai Pustaka on Modern Indonesian Literature," *BSOAS* 35 (1972): 117.

98. James T. Siegel, *A New Criminal Type in Jakarta: Counter-Revolution Today* (Durham, NC: Duke University Press, 1998), 16–17.

99. Michael Bodden, "Dynamics and Tensions of LEKRA's Modern National Theatre, 1959–1965," in *Heirs to World Culture: Being Indonesian, 1950–1965*, ed. Jennifer Lindsay and Maya H. T. Liem (Leiden, the Netherlands: KITLV Press, 2012), 455–58. Also see Pramoedya A. Toer, *Realisme-Sosialis dan Sastra Indonesia* [Social realism and Indonesian literature] (Jakarta: Lentera Dipantara, 2003).

100. Raymond Williams, *Marxism and Literature* (Oxford: Oxford University Press, 1977), 45.

101. Williams explained that literature developed into bourgeois consumption that was read as a part of the creation of taste instead of active actions of writers conditioned by their social reality. Williams, *Marxism and Literature*, 50.

102. Teeuw, "Impact of Balai Pustaka," 116–17.

103. Anderson mistakenly thought Soemantri's *Rasa merdika* was written by Mas Marco, who produced many literary works in the previous years of the *pergerakan* (movement), but Marco was not active in the *pergerakan* from 1920 to 1924 due to a disagreement with the communist Darsono and Semaoen on their attack to Tjokroaminoto. He also continuously faced charges during this time, and secluded himself in the city Salatiga. Anderson, *Imagined Communities*, 32n53.

104. Tjempaka-Pasoeroean, "Aliran djaman atau seorang gadis yang sengsara" [The flow of time or a girl who suffers], *Api*, September 10, 1925.

105. Tjempaka-Pasoeroean, "Aliran djaman atau seorang gadis yang sengsara" [The flow of time or a girl who suffers], *Api*, August 25, 1925.

106. Tjempaka-Pasoeroean, "Aliran djaman atau seorang gadis yang sengsara," [The flow of time or a girl who suffers], *Api*, August 13, 1925.

107. James T. Siegel, in *A New Criminal Type in Jakarta*, reviews the fictional literature during the Indonesian revolution as a struggle for a new kind of kinship model. Also see James Siegel, "The Idea of Indonesia Continues: The Middle Class Ignores Acehnese," *Archipel* 64 (2002), 210–11.

108. In her article "Models and Maniacs: Articulating the Female in Indonesia" in *Fantasizing the Feminine in Indonesia*, ed. Laurie J. Sears (Durham, NC: Duke University Press, 1999), 47–70, Tiwon traces the history of women's articulation and finds that, even when women articulate themselves, they often were still caught in the trap of model and maniac, the "Kartini/Gerwani complex." They can articulate themselves as a female model like Kartini as long as they do not pose a threat to power (like the pre-1965 communist Gerwani that opposed both patriarchy and capitalism). For a different discussion on representation and articulation of women in Indonesian literature, see Tineke Hellwig, *In the Shadow of Change: Images of Women in Indonesian Literature* (Berkeley: Centers for South and Southeast Asia Studies, University of California at Berkeley, 1994).

109. A. Teeuw, "The Impact of Balai Pustaka on Modern Indonesian Literature," *Bulletin of the School of Oriental and African Studies, University of London* 35, no. 1 (1972): 118.

110. Abdoel Moeis, *Salah Asuhan, Cetakan kesembilan* (Jakarta: P. N. Balai Pustaka, 1967).

111. Doris Jedamski, "Balai Pustaka: A Colonial Wolf in Sheep's Clothing," *Archipel* 44 (1992): 38.

112. Jedamski, "Balai Pustaka," 38.

113. Toer, *Sang pemula*, 295.

114. Toer, *Sang pemula*, 301–52.

115. Elsewhere, Gaye Tuchman calls this the "symbolic annihilation of women." Women were either ignored by the media or portrayed in stereotypical roles. Gaye Tuchman, "The Symbolic Annihilation of Women by the Mass Media," in *Culture and Politics*, ed. Lane Crothers and Charles Lockhart (New York: St. Martin's Press, 2000), 150–74.

116. Saskia Wieringa, "IBU or the Beast: Gender Interests in Two Indonesian Women's Organizations," *Feminist Review*, no. 41 (1992): 101–3; Kathryn Robinson, *Gender, Islam and Democracy in Indonesia* (London: Routledge, 2010), 59–60; and Susan Blackburn, *Women and the State in Modern Indonesia* (Cambridge: Cambridge University Press, 2009), 24.

3. GEOGRAPHY OF RESISTANCE

1. Anne Kelly Knowles, Tim Cole, and Alberto Giordano, eds., *Geographies of the Holocaust* (Bloomington: Indiana University Press, 2014); and Anne Kelly Knowles, *Past Time, Past Place: GIS for History* (Redlands, CA: ESRI Press, 2002).

2. Ruth McVey, "Building Behemoth: Indonesian Constructions of the Nation-State," in *Making Indonesia*, ed. Daniel S. Lev and Ruth McVey (Ithaca, NY: Southeast Asia Program Publications, Cornell University, 1996), 13.

3. Benedict Anderson, *Language and Power: Exploring Political Cultures in Indonesia* (Ithaca, NY: Cornell University Press, 1990), 98. Semaoen's report to the Comintern states the state apparatus comprises 97 percent native and 3 percent Dutch who occupied the leading positions. He further says "The lower strata of the intelligentsia are natives." "Brieven van Cominternfunctionarissen aan de PKI, Darsono en Tan Malakka" (Letters from Comintern Officials to the PKI, Darsono and Tan Malaka), July 29, 1926, 9, collection ID: ARCH01744, no. 16, Archief Komintern—Partai Komunis Indonesia, International Institute of Social History (henceforth IISH), Amsterdam.

4. Racial segregation was a part of Dutch colonial system, with Europeans as the highest class, the Chinese as the second, and the indigenous population last. The Chinese class was given greater opportunities and operated their businesses more easily than the natives for the purpose of Dutch commercial interests. Onghokham, "Chinese Capitalism in Dutch Java," *Southeast Asian Studies* 27 (September 1989): 162.

5. For a brief history of the period, see M. C. Ricklefs, *A History of Modern Indonesia since c. 1200* (Stanford, CA: Stanford University Press, 2001 [1991]). For a more comprehensive history of the period, Takashi Shiraishi, *An Age in Motion: Popular Radicalism in Java, 1912–1926* (Ithaca, NY: Cornell University Press, 1990); Ruth T. McVey, *The Rise of Indonesian Communism* (Ithaca, NY: Cornell University Press, 1965); Fritjof Tichelman, *Socialisme in Indonesië: Bronnenpublicatie: De Indische sociaal-democratische vereeniging, 1897–1917* [Socialism in Indonesia: Source publication: The Indische social-democratic association, 1897–1917] (Dordrecht, the Netherlands: Foris, 1985); Emile Schwidder and Tichelman, *Socialisme in Indonesië: Het proces Sneevliet, 1917* [Socialism in Indonesia: The Sneevliet trial, 1917] (Leiden, the Netherlands: KITLV, 1991); A. P. E. Korver, *Sarekat Islam, 1912–1916: Opkomst, bloei en structuur van Indonesie's eerste massabeweging* [Rise, development, and structure of Indonesia's first mass movement] (Amsterdam: Historisch Semenarium van de Universiteit van Amsterdam, 1982); Soe Hok Gie, *Di bawah lentera merah: Riwayat Sarekat Islam Semarang, 1917–1920* [Under the red lantern: The history of Sarekat Islam Semarang, 1917–20] (Yogyakarta, Indonesia: Yayasan Bentang Budaya, 1999); and John T. P. Blumberger, *De communistische beweging in Nederlandsch-Indië* [The communist movement in the Dutch East Indies] (Haarlem, the Netherlands: H. D. Tjeenk Willink & Zoon, 1935).

6. The Russian revolution of 1917 inspired the ISDV so much that they saw it as the path to follow in Indonesia. Sneevliet's article on the Russian revolution titled "Zegepraal" ("Victory") published in *de Indier* blatantly provoked a struggle against colonial rule. Because of this he was arrested, and a year later was sent in exile and banned from coming back to the Indies, leaving behind his wife and their children.

7. The Chinese Communist Party was established in 1921, the Japanese Communist Party was founded in 1922, and in 1930 four communist parties were established in Southeast Asia, including the Indochinese Communist Party, the Siamese Communist Party, the Malayan Communist Party, and the Philippine Communist Party. Onimaru Takeshi, "Living 'Underground' in Shanghai: Noulens and the Shanghai Comintern Network," in *Traveling Nation-Makers: Transnational Flows and Movements in the Making of Modern Southeast Asia*, ed. Caroline S. Hau and Kasian Tejapira (Kyoto: NUS Press in association with Kyoto University Press), 99.

8. Ricklefs, *A History of Modern Indonesia*, 206–26.

9. Anton E. Lucas, *Peristiwa tiga daerah: Revolusi dalam revolusi* [The three regions affair: The revolution inside a revolution] (Jakarta: Pustaka Utama Grafiti, 1989).

10. By 1926, there were no longer reports on OVs.

11. While "intercoder reliability"—a procedure in which two or more coders agree on the coding of content variables to show that the coding scheme is reliable—is typically considered necessary, I proceeded without it during my archival research due to constraints on time and distance. On SPSS Statistics, see William E. Wagner III, *Using SPSS for Social Statistics and Research Methods*, 2nd ed. (London: Pine Forge Press, 2010); and Andy P. Field, *Discovering Statistics Using SPSS (and Sex and Drugs and Rock 'n' Roll)*, 2nd ed. (Thousand Oaks, CA: SAGE Publications, 2005).

12. Robert Cribb, *Historical Atlas of Indonesia* (Honolulu: University of Hawai'i Press, 2000), 126.

13. For example, Gerry van Klinken demonstrates communist activities among Dayak people in Borneo in "Dayak Ethnogenesis and Conservative Politics in Indonesia's Outer Islands" (unpublished manuscript, KITLV, Leiden, the Netherlands, 2001).

14. Shiraishi, *Age in Motion*; and McVey, *Rise of Indonesian Communism*

15. Audrey Kahin, *Rebellion to Integration: West Sumatra and the Indonesian Polity, 1926–1998* (Amsterdam: Amsterdam University Press, 1999); and Steve Farram, "From 'Timor Koepang' to 'Timor NTT': A Political History of West Timor, 1901–1967" (PhD diss., Charles Darwin University, 2004).

16. *Memori serah jabatan 1921–1930 (Jawa Tengah) (Memorie van overgave)* [Memories of the handover of office 1921–30 (Central Java)] (Jakarta: Arsip Nasional Republik Indonesia, 1977), XXXIII–XXXV.

17. All these train and tram companies were headquartered in Semarang, and SCS connected Semarang and Cirebon via Kendal. On this train, Semarang-Batavia could be reached within eight hours. NIS connected Semarang and Surabaya via Gundi and Cepu. On NIS, Semarang-Surabaya could be reached within seven hours. NIS also connected Semarang with Surakarta and Yogya via Kedungjati. Meanwhile, SJS connected Semarang and Rembang via Demak, Kudus, and Pati. *Memori serah jabatan*, XLIX.

18. *Memori serah jabatan*, XXXVII.

19. *Memori serah jabatan*, CLXV.

20. *Memori serah jabatan*, LII–LIX.

21. *Memori serah jabatan*, LXII–LXIII.

22. *Memori serah jabatan*, LXIII.

23. The number of the meetings in the Outer Islands might seem small, but the information channels that could send reports to *Sinar Hindia* headquarter in Semarang, especially from the Outer Islands, were limited, and the practice of reporting was relatively uncommon. To trace if the direction of the expansion changed over time since the early communist movement in 1920 to 1926 through 1965, it is useful to compare maps 1, 2, 3, and 4 with the map of PKI votes in 1957 in Donald Hindley, *The Communist Party of Indonesia, 1951–1963* (Berkeley: University of California Press, 1966), 226.

24. Sugar factories in Java in 1937 showed both existing and closed factories in the past decades. W. A. I. M. Segers, *Changing Economy in Indonesia: A Selection of Statistical Source Material from the Early 19th Century up to 1940*, ed. P. Boomgaard, vol. 8, *Manufacturing Industry 1870–1942* (Amsterdam: Royal Tropical Institute [KIT], 1991), 224.

25. "Brieven van Cominternfunctionarissen aan de PKI, Darsono en Tan Malakka," July 1922, 4-5, Archief Komintern—Partai Komunis Indonesia.

26. McVey, *Rise of Indonesian Communism*, 105.

27. "Document No. 3 the National Front," 1, collection ID: ARCH01744, no. 32, Archief Komintern—Partai Komunis Indonesia, IISH, Amsterdam.

28. Ricklefs, *A History of Modern Indonesia*, 202.

29. McVey, *Rise of Indonesian Communism*, 14, 42, and 111.

30. McVey, *Rise of Indonesian Communism*, 147.

31. "Korban-korban pemogokan" [The Victims of the Strike], "Apakah djadinja" [What happens], "Pergerakan kita" [Our movement], *Sinar Hindia*, May 12, 1923.

32. McVey, *Rise of Indonesian Communism*, 148.

33. "Apakah Djadinja."

34. "Perampoean bergerak" [Women in motion], *Sinar Hindia*, September 11, 1918; SAM, "Semaoen," *Sinar Hindia*, May 29, 1923.

35. "Brieven van Cominternfunctionarissen aan de PKI, Darsono en Tan Malakka."

36. "Brieven van Cominternfunctionarissen aan de PKI, Darsono en Tan Malakka."

37. "Document No. 3 the National Front."

38. "Document No. 3 the National Front."

39. "Document No. 3 the National Front."

40. "Document No. 3 the National Front."

41. "Brieven van Cominternfunctionarissen aan de PKI, Darsono en Tan Malakka."

42. "Notulen van vergadering van het subsecretariaat voor India en Indonesië" [Minutes of the meeting of the Sub-Secretariat for India and Indonesia], collection ID: ARCH01744, no. 2, Archief Komintern—Partai Komunis Indonesia, IISH, Amsterdam. In this map, Giham and Menggala are combined under Lampung.

43. "Brieven van Cominternfunctionarissen aan de PKI, Darsono en Tan Malakka," Archief Komintern—Partai Komunis Indonesia.

44. The Dutch police numbered 1,249, while the native police numbered 25,000. The Dutch army was all together 33,000, of which the Europeans were 7,363 and the natives were 26,000. In the navy, there were 1,633 Europeans and 1,163 natives. Of the state employees, 20,000 were Dutch, while 139,927 were natives. "Komintern – Partai Komunis Indonesia" archive, inventory number 16, 54.

45. See *Sinar Hindia*, May 8, 1921.

46. "Brieven van Cominternfunctionarissen aan de PKI, Darsono en Tan Malakka."

47. "Brieven van Cominternfunctionarissen aan de PKI, Darsono en Tan Malakka."

48. Leo Suryadinata, "Pre-War Indonesian Nationalism and the Peranakan Chinese," *Indonesia*, no. 11 (April 1971).

49. See *Api*, January 29, August 7, and September 19, 1925.

50. "Nasib Tahoen 1925" [The fate of the year 1925], *Api*, January 2, 1926.

51. For more on *stamboel*, see Matthew Isaac Cohen, *Komedie Stamboel: Popular Theater in Colonial Indonesia, 1891–1903* (Athens: Ohio University Press, 2006). On cinema, see Dafna Ruppin, *The Komedi Bioscoop: The Emergence of Movie-Going in Colonial Indonesia, 1896–1914* (Bloomington: Indiana University Press, 2016).

4. *OPENBARE VERGADERINGEN* AND CULTURES OF RESISTANCE

1. John T. P. Blumberger, *De communistische beweging in Nederlandsch-Indië* [The communist movement in the Dutch East Indies] (Haarlem, Netherlands: H. D. Tjeenk

Willink & Zoon, 1935); John Ingelson, *Workers, Unions and Politics: Indonesia in the 1920s and 1930s* (Leiden: Brill, 2014); A. P. E. Korver, *Sarekat Islam, 1912–1916: Opkomst, bloei en structuur van Indonesie's eerste massabeweging* [Sarekat Islam, 1912–1916: Rise, development and structure of Indonesia's first mass movement] (Amsterdam: Historisch Semenarium van de Universiteit van Amsterdam, 1982); Ruth T. McVey, *The Rise of Indonesian Communism* (Ithaca, NY: Cornell University Press, 1965); Takashi Shiraishi, *An Age in Motion: Popular Radicalism in Java, 1912–1926* (Ithaca, NY: Cornell University Press, 1990); and Fritjof Tichelman, *Socialisme in Indonesië: Bronnenpublicatie: De Indische sociaal-democratische vereeniging, 1897–1917* [Socialism in Indonesia: Source publication: The Indische social-democratic association, 1897–1917] (Dordrecht, Netherlands: Foris, 1985).

2. Nicholas Tarling, *Nationalism in Southeast Asia: 'If the People Are with Us'* (New York: Routledge, 2012), 109.

3. R. M. Soewardi Soerjaningrat, *Als ik eens Nederlander was,. . .* [If I were a Dutch man,. . .], trans. R. M. Soewardi Soerjaningrat (Bandoeng: Comité Boemipoetra, 1913), 11–14. Author's translation.

4. James T. Siegel, *Fetish, Recognition, Revolution* (Princeton, NJ: Princeton University Press, 1997), 26.

5. Siegel, *Fetish, Recognition, Revolution*, 27. See the Malay version in Soerjaningrat, *Als ik eens Nederlander was,. . .*

6. "Propaganda vergadering Serikat Ra'jat" [Propaganda meeting of People's Union], *Sinar Hindia*, July 24, 1924.

7. Reconstructed from various descriptions of OVs reported in *Sinar Hindia*.

8. The arts have been important aspects of social movements. Some examples are songs in French revolution and literature, music, opera, and films in the United States in 1930s. Laura Mason, *Singing the French Revolution: Popular Culture and Politics, 1787–1799* (Ithaca, NY: Cornell University Press, 1996); and Michael Denning, *The Cultural Front: The Laboring of American Culture in the Twentieth Century* (London: Verso, 2010).

9. On songs and politics in Indonesia, see Steven Farram, "'Ganyang!' Indonesian Popular Songs from the Confrontation Era, 1963–1966," *Bijdragen tot de Taal-, Land- en Volkenkunde* 170, no. 1 (2014): 1–24; Harry Poeze, "Songs as a Weapon" (unpublished manuscript, the Royal Netherlands Institute of Southeast Asian and Caribbean Studies (KITLV), Leiden, the Netherlands, 2013).

10. Ramelan, "Toedjoeh atau enam djam?" [Seven or six hours?], *Sinar Hindia*, April 29, 1924; "Enam djam bekerdja" [Six working hours], *Sinar Hindia*, April 26, 1924.

11. Some lyrics were direct translations of the original: Hendrik de Man, *The Psychology of Marxian Socialism* (New Brunswick, NJ: Transaction Books, 1985 [1928]), 158; Carl Strikwerda, *A House Divided: Catholics, Socialists, and Flemish Nationalists in Nineteenth-Century Belgium* (Lanham: Rowman & Littlefield, 1997), 115.

12. Lynn Hunt, *Politics, Culture, and Class in the French Revolution* (Berkeley: University of California Press, 2004 [1984]), 62–65, 86.

13. de Man, *The Psychology of Marxian Socialism*, 158.

14. "Satoe Mei di Semarang" [May First in Semarang], *Sinar Hindia*, May 2, 1924.

15. Maurice Agulhon, *Marianne into Battle: Republican Imagery and Symbolism in France: 1789–1880* (Cambridge: Cambridge University Press, 1981), 2.

16. Jan Mrázek, *Phenomenology of a Puppet Theatre: Contemplations on the Art of Javanese Wayang Kulit* (Leiden, the Netherlands: KITLV, 2005); Edward C. Van Ness and Shita Prawirohardjo, *Javanese Wayang Kulit: An Introduction* (Kuala Lumpur: Oxford University Press, 1980); Carl J. Hefner, "Ludruk Folk Theatre of East Java: Toward a Theory of Symbolic Action" (PhD diss., University of Hawai'i, 1994); Barbara Hatley, *Performing Contemporary Indonesia: Celebrating Identity, Constructing Community* (Leiden, the

Netherlands: Brill, 2015); Matthew Cohen, *Inventing the Performing Arts: Modernity and Tradition in Colonial Indonesia* (Honolulu: University of Hawai'i Press, 2016).

17. "Perayaan 1 Mei" [May First celebration], *Api*, May 5, 1925.

18. "De geinterneerden" [The internees], *De locomotief,* February 8, 1927; "Oprichtingsvergadering van den bond van drukkerijpersoneel" [Founding meeting of printing personnel], *Algemeen Handelsblad voor Nederlandsch-Indie,* June 16, 1924. Obviously, it is not clear if she was of a *prijaji* class—it was common for local people to save up by keeping gold jewelry.

19. Barbara Hatley, *Javanese Performances on an Indonesian Stage: Contesting Culture, Embracing Change* (Honolulu: University of Hawai'i Press, 2008), 9–12.

20. "Derma pergerakan" [Charity of the movement], *Sinar Hindia,* May 22, 1920; "S.I. Salatiga" [SI Salatiga], December 31, 1921; "Loedroek, Schoolfuil dan 7 Nov. Sovjet Roeslan oleh S.I. School Ngandjoek" [Loedroek, Schoolfuil, and (the celebration of) the November 7 Soviet Russia by SI School Ngandjoek], *Sinar Hindia,* November 14, 1923; and "Peraja'an Serikat Islam Salatiga" [The Celebration of Sarekat Islam Salatiga], *Sinar Hindia,* January 7, 1924.

21. M. C. Ricklefs, *A History of Modern Indonesia since c.1300,* 2nd ed. (London: MacMillan, 1991).

22. "Vergadering van de Sarekat Islam over Indië Weerbaar te Moearatewe" [Meeting of the Sarekat Islam about the Indies' resilience in Moearatewe], code 86968, album number 478, October 15, 1916, KITLV Digital Image Library, http://hdl.handle.net/1887.1/item:914752.

23. See advertisements in *Sinar Hindia,* January 14, 1919, March 29, 1922, and February 13, 1924.

24. Peter Boomgaard, ed., *Changing Economy in Indonesia: A Selection of Statistical Source Material from the Early 19th Century up to 1940,* vol. 13, *Wages 1820–1940* (Amsterdam: Royal Tropical Institute [KIT], 1992), 140.

25. James Van Horn Melton, *The Rise of the Public in Enlightenment Europe* (New York: Cambridge University Press, 2001), 197–199.

26. Melton, *The Rise of the Public,* 203 and 274.

27. Quoted from W. M. F. Mansvelt, "Onderwijs en communisme" [Education and communism], *Koloniale Studien* 12 (1928), 202–25, in Fred R. von der Mehden, "Marxism and Early Indonesian Islamic Nationalism," *Political Science Quarterly* 73, no. 3 (September 1958), 337n1.

28. *Sinar Hindia,* July 31, 1924. Italics in original.

29. "Verslag pendek S.I. Wadon Ambarawa" [A short report on S.I. Wadon Ambarawa], *Sinar Hindia,* April 27, 1920.

30. Also see Ruth McVey, "Teaching Modernity: The PKI as an Educational Institution," *Indonesia,* no. 50 (1990): 5–27; and Ruth T. McVey, "Taman Siswa and the Indonesian National Awakening," *Indonesia,* no. 4 (1967): 128–49. For other research on education in Indonesia, see: Agus Suwignyo, "The Great Depression and the Changing Trajectory of Public Education Policy in Indonesia, 1930–42," *Journal of Southeast Asian Studies* 44, no. 3 (2013): 465–89; Lee Kam Hing, "Taman Siswa in Postwar Indonesia," *Indonesia,* no. 25 (April, 1978): 41–59; and I. J. Brugmans, *Geschiedenis van het onderwijs in Nederlandsch-Indie* [The history of the education in the Dutch East Indies] (Groningen, the Netherlands: J. B. Wolters Uitgevers Maatschappij, 1938).

31. "Notulen van vergadering van het subsecretariaat voor India en Indonesië" [Minutes of the meeting of the Sub-Secretariat for India and Indonesia], collection ID: ARCH01744, no. 2, Archief Komintern—Partai Komunis Indonesia, International Institute of Social History (henceforth IISH), Amsterdam.

32. "Pendapatan kami terhadap Inlandsch Onderwijs Kongres" [Our opinion on the Inlandsch Onderwijs (Dutch: Native Education) Congress], *Sinar Hindia*, May 1, 1922.

33. William Frederick notes that one quarter or more of the teaching staff in many non-Dutch schools was of a *prijaji* rank. William Frederick, *Visions and Heat: The Making of the Indonesian Revolution* (Athens: Ohio University Press, 1988), 61. See also McVey, "Taman Siswa and the Indonesian National Awakening"; Kenji Tsuchiya, *Democracy and Leadership: The Rise of the Taman Siswa Movement in Indonesia* (Honolulu: University of Hawai'i Press, 1987).

34. Tan Malaka, *S.I. Semarang dan Onderwijs* [SI Semarang and education] (Semarang, Indonesia: SI School, 1921), 16, microfilm collection, IISH, Amsterdam.

35. "Hidoeplah S.I. scholen!" [Long live SI schools!], *Sinar Hindia,* January 23, 1924.

36. "Njonjah Sneevliet, selamat-djalan!" [Mrs. Sneevliet, farewell!], *Sinar Hindia*, July 4, 1923. The website of the IISH in the Netherlands—based on the primary documents housed there—notes that Sneevliet was married four times. He and Brouwer were divorced in 1924 soon after Brouwer returned to the Netherlands in 1923. Source: IISH, "Henk Sneevliet – A Life in Documents," accessed December 12, 2018, http://www.iisg.nl/collections/sneevliet/life-8.php.

37. Frederick, *Visions and Heat*, 59.

38. For Tan Malaka's biography, see Harry A. Poeze, *Tan Malaka: Strijder voor Indonesië's Vrijheid: Levensloop van 1897 tot 1945* [Tan Malaka: Fighter for Indonesia's freedom, life history from 1897 to 1945] ('s-Gravenhage: Nijhoff, 1976); and *Verguisd en vergeten: Tan Malaka, de linkse beweging en de Indonesische Revolutie, 1945–1949* [*Maligned and forgotten: Tan Malaka, the left movement and the Indonesian revolution, 1945-1949*], 3 vols. (Leiden, the Netherlands: KITLV Uitgeverij, 2007).

39. Tan Malaka, *Indoneziya i yeye mesto na probuzhdayushchemsya vostoke* [Indonesia and its place in the awakening East] (Moscow: State Publishing House, 1925), 135.

40. Tan Malaka, *S.I. Semarang dan Onderwijs*, 14.

41. Cf. "Korban-Korban PARI" [The victims from PARI], August 1933, 14, "Komintern – Partai Komunis Indonesia" archive, inventory number 29, IISH, Amsterdam.

42. Audrey Kahin, *Rebellion to Integration: West Sumatra and the Indonesian Polity, 1926–1998* (Amsterdam: Amsterdam University Press, 1999).

43. *Sinar Hindia*, December 15, 1920.

44. "Geraknja S.I. Perampoean Semarang" [The movement of SI Women Semarang], *Sinar Hindia*, December 20, 1920.

45. "Snel School Salatiga" [Quick School Salatiga], *Sinar Hindia*, June 29, 1922.

46. "S.I. Ngandjoek Dengan Woro Roessilah Hadisoebroto" [SI Ngandjoek with Ms. Roessilah Hadisoebroto], *Sinar Hindia*, August 1, 1921.

47. *Sinar Hindia,* July 16, 1924.

48. This is a sample invitation to an OV from *Sinar Hindia*, January 31, 1924.

49. In 1928, the young Sukarno, who later became the first president of Indonesia, wrote a book *NASAKOM* (*Nasionalisme, agama, komunis* [Nationalism, religion, communism]) that tried to unite the seeming differences among these three ideologies. Soekarno and Ruth Thomas McVey, *Nationalism, Islam, and Marxism* (Ithaca, NY: Modern Indonesia Project, Southeast Asia Program, Cornell University, 1970). Tjokroaminoto also wrote a book on socialism and Islam: *Islam dan socialisme* [Islam and socialism] (Jakarta: Penerbit Bulan Bintang, 2003 [1924]).

50. Shiraishi, *Age in Motion*, 249.

51. Darsono, "Kommunisma dan Islamisma" [Communism and Islamism] [part 1], *Sinar Hindia,*

February 14, 1921; and "Kommunisma dan Islamisma" [Communism and Islamism] [part 2], *Sinar Hindia*, February 15, 1921; "Kommunisma dan Islamisma" [Communism and Islamism] [part 3], *Sinar Hindia*, February 17, 1921.

52. The idea of a "red hajj" (hajj: Muslim who had done pilgrimage to Mecca) became popular.

53. Nor Hiqmah, *H.M. Misbach Kisah Hadji Merah* [H. M. Misbach, the story of a red hajj] (Depok, Indonesia: Komunitas Bambu, 2008), 77. Author's translation.

54. For his view on "Islamic communists," see Shiraishi, *Age in Motion*, 249.

55. See Kahin, *Rebellion to Integration*; Steve Farram, "From 'Timor Koepang' to 'Timor NTT': A Political History of West Timor, 1901–1967" (PhD diss., Charles Darwin University, 2004); and Steve Farram, "Revolution, Religion and Magic: The PKI in West Timor, 1924–1966," *Bijdragen tot de Taal-, Land- En Volkenkunde* 158, no. 1 (2002), 21–48.

56. Tjempaka-Pasoeroean, "Aliran djaman atau seorang gadis yang sengsara" [The flow of time or a girl who suffers], *Api*, September 1, 1925.

57. Elsbeth Locher-Scholten, *Women and the Colonial State: Essays on Gender and Modernity in the Netherlands Indies 1900–1942* (Amsterdam: Amsterdam University Press), 188–90.

58. Djoeinah, "Pemimpin istri dalam Vergadering (Oengaran)" [Women's leaders in *vergadering* (Oengaran)], *Sinar Hindia*, October 2, 1920.

59. D. T., "Terhadap kaoem poeteri terpeladjar" [To the educated women], *Sinar Hindia*, February 16, 1921.

60. Women's involvement beyond the Indies can be traced in the story of an exiled communist member Soekaesih who was found in the Netherlands after the revolt 1926 to 1927. From the Politieke Inlichtingendienst (the Political Intelligence Service, PID) interview in 's-Gravenhage on October 27, 1937, "Communisten verbannen uit Indië" [Communists expelled from the Indies], no. 52450, *Rapporten Centrale Inlichtingendienst, 1919-1940*, historici.nl now in Huygens Instituut, Amsterdam, https://resources. huygens.knaw.nl/rapportencentraleinlichtingendienst/data/IndexResultaten/IndexVen sterResultaat?persoon=Philippo-Soekaesih%20(Soelasih),%20Raden. On the involvement of women in the communist movement in postindependence Indonesia, see: Saskia Wieringa, *Sexual Politics in Indonesia* (New York: Palgrave Macmillan, 2002); and Annie Pohlman, "Janda PKI: Stigma and Sexual Violence against Communist Widows Following the 1965–1966 Massacres in Indonesia," *Indonesia and the Malay World* 44, no. 128 (2016): 68–83.

61. E. M. Uhlenbeck, "Systematic Features of Javanese Personal Names," *Word* 25, nos. 1–3 (1969): 332–333.

62. "'Brief aus Indonesien,' Met Nederlandse vertaling" ["'Letter from Indonesia,' with Dutch translation"], collection ID: ARCH01744, no. 15, Archief Komintern—Partai Komunis Indonesia, IISH, Amsterdam.

5. "THE WAR OF PENS AND WORDS"

1. A version of this chapter has previously appeared as Rianne Subijanto, "Enlightenment and the Revolutionary Press in Colonial Indonesia," *International Journal of Communication* 11 (2017): 1357–77, and translated to Dutch as Rianne Subijanto, "Verlichtingsidealen en de revolutionaire pers in koloniaal Indonesië: De inhoud, productie en distributie van *Sinar Hindia*," *Tijdschrift voor Geschiedenis* 130, no. 3 (2017): 449–66.

2. Soekin, "Pergantian Hawa Dari Tahoen 1921 ke 1922" [The change of atmosphere from 1921 to 1922], *Sinar Hindia*, January 2, 1922.

3. Ahmat Adam, *The Vernacular Press and the Emergence of Modern Indonesian Consciousness (1855–1913)* (Ithaca, NY: Southeast Asia Program, Cornell University, 1995).

Also see Robert. E. Elson, *The Idea of Indonesia: A History* (Cambridge, UK: Cambridge University Press, 2008).

4. Benedict Anderson, *Imagined Communities: Reflections on the Origin and Spread of Nationalism* (London: Verso, 1991), 67.

5. John T. P. Blumberger, *De communistische beweging in Nederlandsch-Indië* [The communist movement in the Dutch East Indies] (Haarlem, the Netherlands: H. D. Tjeenk Willink & Zoon, 1935); A. P. E. Korver, *Sarekat Islam, 1912–1916: Opkomst, bloei en structuur van Indonesie's eerste massabeweging* [Sarekat Islam, 1912–1916: Rise, development and structure of Indonesia's first mass movement] (Amsterdam: Historisch Semenarium van de Universiteit van Amsterdam, 1982); Ruth T. McVey, *The Rise of Indonesian Communism* (Ithaca, NY: Cornell University Press, 1965); Takashi Shiraishi, *An Age in Motion: Popular Radicalism in Java, 1912–1926* (Ithaca, NY: Cornell University Press, 1990); Fritjof Tichelman, *Socialisme in Indonesië: Bronnenpublicatie: De Indische sociaaldemocratische vereeniging, 1897–1917* [*Socialism in Indonesia: Source publication: The Indische social-democratic association, 1897–1917*] (Dordrecht, the Netherlands: Foris, 1985); and Benedict R. O'G. Anderson, "Language, Fantasy, Revolution: Java 1900–1950," in *Making Indonesia*, ed. Daniel S. Lev and Ruth T. McVey (Ithaca, NY: Cornell University Press, 1996), 26–40.

6. Wolfgang Behn, "Revolutionary Publishing in Iran from the Overthrow of the Shah until His Death," *MELA Notes* 21 (1980): 11–15; Palmira Brummett, *Image and Imperialism in the Ottoman Revolutionary Press: 1908–1911* (Albany: SUNY Press, 2000); Charles Cutler, *Connecticut's Revolutionary Press* (Chester, CT: Pequot Press, 1975); Patrick Daley, "Newspaper Competition and Public Spheres in New Hampshire in the Early Revolutionary Period," *Journalism & Communication Monographs* 11, no. 1 (2009): 3–65; Melvin Edelstein, "La Feuille Villageoise, the Revolutionary Press, and the Question of Rural Political Participation," *French Historical Studies* 7, no. 2 (1971): 175–203; and Jeremy Popkin, *Revolutionary News: The Press in France 1789–1799* (Durham, NC: Duke University Press, 1999).

7. Adam, *Vernacular Press*, 16.

8. Adam, *Vernacular Press*, 108, 71–78.

9. Adam, *Vernacular Press*, 110–11; M. C. Ricklefs, *A History of Modern Indonesia Since C. 1200* (Stanford, CA: Stanford University Press, 2001), 207–8; and Shiraishi, *Age in Motion*, 51, 58.

10. Robert Cribb, *Historical Atlas of Indonesia* (Honolulu: University of Hawai'i Press, 2000).

11. Shiraishi, *Age in Motion*, 245.

12. Anderson, "Language, Fantasy, Revolution," 35, 37.

13. Anderson, "Language, Fantasy, Revolution," 35–36.

14. Synthema [Soemantri], "Kemerdika'an soeara pers. Menaboer benih kebentjian?" [The freedom of the voice of the press. Sowing the seeds of hatred?], *Sinar Hindia*, April 9, 1924.

15. G. W. J. Drewes, "D. A. Rinkes: A Note on His Life and Work," *Bijdragen tot de Taal-, Land- en Volkenkunde* 117, no. 4 (1961): 417–35.

16. Herlambang Wiratraman, "Press Freedom, Law and Politics in Indonesia: A Social-Legal Study" (PhD diss., Leiden University, 2014), 52–53.

17. With the price of a single copy of the newspaper at 0.10 *roepijah*, the proposed fine was heavy. *Roepijah* and gulden were both the circulating units of currency bearing the same value (e.g., 1 *roepiah* = 1 gulden). See Roger Lane, *Encyclopedia of Small Silver Coins*, 3rd ed. (Raleigh: Lulu Press, 2008), 382–84. Guilders was another term used interchangeably with them.

18. Synthema, "Kemerdika'an soeara pers."

19. Mirjam Maters, *Dari perintah halus ke tindakan keras: Pers zaman kolonial antara kebebasan dan pemberangusan, 1906–1942* (Jakarta: KITLV, 2003), 136.

20. "Tegen het voortwoekerend communisme" [Against propagating communism], *De Sumatra Post,* April 14, 1924.

21. Pramoedya Ananta Toer, *Sang pemula* (Jakarta: Hasta Mitra, 1985), 157–58, 161.

22. Synthema, "Kemerdika'an soeara pers"

23. Jurgen Habermas, "Further Reflections on the Public Sphere," in *Habermas and the Public Sphere,* ed. C. Calhoun (Cambridge, MA: MIT Press, 1992), 440.

24. D. A. S., "Persdelict dan preventief" [Press offense and preemptive], *Api,* October 2, 1925.

25. Koetoe Bolspik, "Djournalist tjingtjau" [Journalists of mushrooms], *Api,* August 21, 1925.

26. D. A. S., "Persdelict dan preventief."

27. "Journalistiek Indonesia" [Indonesian journalism], *Api,* October 26, 1925.

28. D. A. S., "Persdelict dan preventief."

29. Joshua Atkinson, *Alternative Media and Politics of Resistance: A Communication Perspective* (New York: Peter Lang, 2010), 18.

30. Shiraishi, *Age in Motion,* 48, 59. In the contemporary context in which the word "propagandist" carries a negative meaning, it is probably more fitting to call them instead "activists-turned-journalists."

31. Dewi Yuliati, *Semaoen, pers bumiputera, dan radikalisasi Sarekat Islam Semarang* [Semaoen, native press, and the radicalization of Sarekat Islam Semarang] (Semarang, Indonesia: Penerbit Bendera, 2000), 5; Ricklefs, *History of Modern Indonesia,* 219; and Shiraishi, *Age in Motion,* 53, 81.

32. Cf. Antonio Gramsci, *Selections from the Prison Notebooks* (New York: International Publishers, 1971).

33. Recent research reveals that the early communist movement also reached, besides Java, Aceh, Timor and Alor, Kalimantan, and West Sumatra. Steve Farram, "From 'Timor Koepang' to 'Timor NTT': A Political History of West Timor, 1901–1967" (PhD diss., Charles Darwin University, 2004); Gerry Klinken, "Dayak Ethnogenesis and Conservative Politics in Indonesia's Outer Islands" (unpublished paper, KITLV, Leiden, 2001); and Audrey Kahin, *Rebellion to Integration: West Sumatra and the Indonesian Polity, 1926–1998* (Amsterdam: Amsterdam University Press, 1999).

34. Samuel Zwemer, "The Native Press of the Dutch East Indies," *The Muslim World* 13, no. 1 (1923): 39–49.

35. See also David T. Hill, *The Press in New Order Indonesia* (Nedlands, Australia: University of Western Australia Press, 1994), 14, 29.

36. Adam, *Vernacular Press,* 172, 120.

37. Although both used the word "light" in their names, the practices of the enlightenment project were purposefully adopted only during the time of *SH/Api.*

38. "Journalistiek Indonesia."

39. Administrasi dagblad Api Semarang, "Bagaimana Api bisa dimadjoekan?" [How to develop *Api?*], *Api,* July 30, 1925.

40. Administrasi dagblad Api Semarang, "Bagaimana Api bisa dimadjoekan?"

41. Directie Api, "Apa sebabkah Api begitoe mahal harganja?" [Why is the cost of *Api* expensive?], *Api,* May 28, 1925.

42. Directie Api, "Apa sebabkah Api begitoe mahal harganja?"

43. Redactie, "1 Mei" [May 1], *Sinar Hindia,* May 1, 1919.

44. Toer, *Sang Pemula*; and Zwemer, "Native Press of the Dutch East Indies," 41.

45. "5 kewadjiban bagi abonne Api" [Five duties of Api subscribers], *Api,* March 23, 1926.

46. Directie Api, "Apa sebabkah Api begitoe mahal harganja?"

47. On the history of printing and typography, see Warren Chappell and Robert Bringhurst, *A Short History of the Printed Word* (Point Roberts, WA: Hartley & Marks Publishers, 1999); Daniel Berkeley Updike, *Printing Types: Their History, Forms, and Use* (London: Oak Knoll Press, 2001); and Robert Bringhurst, *The Elements of Typographic Style* (Vancouver: Hartley & Marks, 2019).

48. Philip B. Meggs, *A History of Graphic Design*, 3rd ed. (Hoboken, NJ: John Wiley & Sons, 1998), 147.

49. Directie Api, "Apa sebabkah Api begitoe mahal harganja?"

50. Since Gutenberg revolution in 1440s up until the invention of linotype and monotype in nineteenth century, printing presses used movable types to compose texts to print.

51. Balai Poestaka, *Bureau voor de Volkslectuur. The Bureau of Popular Literature of Netherlands India. What It Is and What It Does*, ed. B. Th. Brondgeest and G. W. J. Drewes (Batavia: Kantoor voor de Volkslectuur, 1930), 24; for another work on Balai Poestaka, see: A. Teeuw, "The Impact of Balai Pustaka on Modern Indonesian Literature," *Bulletin of the School of Oriental and African Studies* 35, no. 1 (1972): 111–27.

52. Shiraishi, *Age in Motion*, 245.

53. The meaning of "*penerangan*" (root: "*terang*," meaning "clear, bright") has evolved over time. Compare Suharto's (1966–98) Ministry of Penerangan ("information"), the regime's propaganda arm. I translate "*penerangan*" as "enlightenment" for the period 1920 to 1926, which better reflects the practices of the revolutionary press.

54. Neil Postman, *Amusing Ourselves to Death: Public Discourse in the Age of Show Business* (New York: Penguin Books, 2006).

55. Marco, "Nasib kami" [Our fate], *Sinar Hindia*, July 11, 1921.

56. Darsono, "Pimpinan, Centraal Sarekat Islam (kritik terhadap kepada Kongres 1920)" [Leadership, Central Sarekat Islam (A critique of the 1920 Congress)] [part 1], *Sinar Hindia*, October 6, 1920; Darsono, "Pimpinan, Centraal Sarekat Islam (kritik terhadap kepada Kongres 1920)" [Leadership, Central Sarekat Islam (A Critique of the 1920 Congress)] [part 2], *Sinar Hindia*, October 7, 1920; and Darsono, "Pimpinan, Centraal Sarekat Islam (kritik terhadap kepada Kongres 1920)" [Leadership, Central Sarekat Islam (A critique of the 1920 Congress)] [part 3], *Sinar Hindia*, October 9, 1920.

57. These are important references that were likely popular at the time as the global Spanish flu pandemic concluded just a few years prior in 1918.

58. Tjokroaminoto, "Pemboekaan rahasia pemimpin C.S.I" [Revealing the secret of CSI leadership], *Sinar Hindia*, October 21, 1920.

59. John Downing, *Radical Media: Rebellious Communication and Social Movements* (Thousand Oaks, CA: SAGE Publications, 2001), 28.

60. McVey, *The Rise of Indonesian Communism*, 61; also see Ben Fowkes and Bulent Gökay, *Muslims and Communists in Post-Transition States* (London, UK: Routledge, 2011).

61. Cribb, *Historical Atlas of Indonesia*, 40.

62. *Sinar Hindia*, March 24, 1924. Italics in the original.

63. Toer, *Sang Pemula*.

64. Anderson, *Imagined Communities*.

65. Charles Tilly, *Popular Contention in Great Britain, 1758–1834* (Boulder, CO: Paradigm, 1995).

6. IN THE NAME OF PUBLIC PEACE AND ORDER

1. "Spreekdelikt jang pertama di Indonesia bagi pemimpin perempoean S. Woro ATI (Padmosoemadhio) dapat gratifikasi satoe tahoen pendjara" [The first speech offense in

Indonesia for women's leader Ms. Ati (Padmosoemadhio) received a one-year sentence], *Api,* January 29, 1926. Author's translation.

2. Audrey R. Kahin, "The 1927 Communist Uprising in Sumatra: A Reappraisal," *Indonesia* 62 (Oct 1996): 19–20; and Michael C. Williams, *Sickle and Crescent: The Communist Revolt of 1926 in Banten* (Ithaca, NY: Cornell Modern Indonesia Project, Cornell University, 1982), 28, 36.

3. Harry J. Benda and Ruth T. McVey, *The Communist Uprisings of 1926–1927 in Indonesia: Key Documents* (Ithaca, NY: Modern Indonesia Project, Cornell University, 1960), xiv. Also see Kahin, "The 1927 Communist Uprising in Sumatra"; Williams, *Sickle and Crescent*; and Sartono Kartodirdjo, *Protest Movements in Rural Java: A Study of Agrarian Unrest in the Nineteenth and early Twentieth Centuries* (London: Oxford University Press, 1973).

4. Takashi Shiraishi, "The Phantom World of Digoel," *Indonesia,* no. 61 (Apr 1996): 116, 118. Also see the discussion on *zaman normal* (normal period) in Takashi Shiraishi, *The Phantom World of Digul* (Singapore: NUS Press, 2021), 20–22.

5. Elsbeth Locher Scholten, *Ethiek in fragmenten: Vijf studies over koloniaal denken en doen van Nederlands in de Indonesische archipel 1877–1942* [Ethics in fragments: Five studies on colonial Dutch thought and action in the Indonesian archipelago 1877–1942] (Utrecht, the Netherlands: Utrecht Hess Publishers, 1981), 176–208, quoted in Ulbe Bosma, *Sugar Plantation in India and Indonesia: Industrial Production, 1770–2010* (Cambridge: Cambridge University Press, 2013), 177n53. On the Ethical Policy in the Indies, see Marieke Bloembergen and R. Raben, *Het koloniale beschavingsoffensief: Wegen naar het nieuwe Indië, 1890–1950* [The colonial civilization offensive: Roads to the new Indies, 1890–1950] (Leiden: KITLV Uitgeverij, 2009).

6. Mirjam Maters, *Dari perintah halus ke tindakan keras: Pers zaman kolonial antara kebebasan dan pemberangusan 1906–1942* [From soft orders to harsh measures: The colonial press between freedom and suppression 1906–1942], trans. Mien Joebhaar (Jakarta: Hasta Mitra & KITLV, 1998), 113.

7. Maters, *Dari perintah halus ke tindakan keras.*

8. Maters, *Dari perintah halus ke tindakan keras,* 311.

9. Maters, *Dari perintah halus ke tindakan keras,* 287.

10. Other research on public order: Chris Greer and Eugene McLaughlin, "We Predict a Riot?: Public Order Policing, New Media Environments and the Rise of the Citizen Journalist," *British Journal of Criminology* 50 (2010): 1041–59; and P. A. J. Waddington, *Liberty and Order: Public Order Policing in a Capital City* (London: UCL Press Limited, 1994).

11. Philippo-Raden Soekaesih and G. van Munster, *Indonesia, een politiestaat* [Indonesia, a police State] (Amsterdam: De Schijnwerper, 1938), collection ID IISG Bro N 477/26 A, International Institute of Social History (IISH), Amsterdam.

12. Benda and McVey, *Communist Uprisings of 1926–1927,* xii, xvi.

13. Maters, *Dari perintah halus ke tindakan keras,* 115.

14. Maters, *Dari perintah halus ke tindakan keras,* 115–16.

15. Maters, *Dari perintah halus ke tindakan keras,* 128–29.

16. Herlambang Perdana Wiratraman, "Press Freedom, Law and Politics in Indonesia: A Socio-Legal Study" (PhD diss., University of Leiden, 2014), 52–53.

17. Maters, *Dari perintah halus ke tindakan keras,* 130–31. Author's translation.

18. Koninklijk Besluit, January 7, 1914, no. 28, *Staatsblad van Nederlandsch-Indie 1914* no. 205, 206 dan 207. The articles on hate speech were articles 63a and b of *Europese wetboek van strafrecht* (*The Book of Criminal Law for Europeans*) and articles 66a and b of *Inlandsche wetboek van strafrecht* (*The Book of Criminal Law for Natives*) implemented on March 15, 1914. In 1918, the *haatzaai artikelen* included articles 154 through 157 of the *Book of Criminal Law*, "Hate speech against government is punishable with fine maximum

of 300 gulden or with imprisonment maximum seven years for all citizens" (Maters, *Dari perintah halus ke tindakan keras*, 130n46 and 131. Author's translation.).

19. Alinoeso, "The Methods of the White Terror to Suppress the Revolutionary Movement in Indonesia," sent from Soerabaya, November 1, 1926, in "Brieven van de PKI Weltevreden aan de CPH" ("Letters from the PKI Weltevreden to the CPH"), collection ID: ARCH01744, no. 20, Archief Komintern—Partai Komunis Indonesia, International Institute of Social History (IISH), Amsterdam. At the time of its publication, Alimin and Moeso had already been in exile abroad, but it is possible that the document was prepared using reports from these two communist leaders.

20. Alinoeso, "Methods of the White Terror," 1–2.

21. All city names are spelled following the document Alinoeso, "Methods of the White Terror." The document also mentions "Pulu Belu Island." It is not clear if this refers to Belu located in the Residentie of Timor en Onderhoorigheden.

22. On anticommunist reactions, also see Ruth T. McVey, *The Rise of Indonesian Communism* (Ithaca, NY: Cornell University Press, 1965), 295; and Takashi Shiraishi, *An Age in Motion: Popular Radicalism in Java, 1912–1926* (Ithaca, NY: Cornell University Press, 1990).

23. Alinoeso, "Methods of the White Terror," 2–3.

24. Alinoeso, "Methods of the White Terror," 3.

25. Alinoeso, "Methods of the White Terror," 3.

26. W. F. Wertheim, *Indonesian Society in Transition: A Study of Social Change* (The Hague: W. van Hoeve Ltd, 1964), 250.

27. "Si manis kontra si merah: Kapital goela setengah mati" ["The sweet versus the red: The capitalist sugar is dying"], *Api*, January 29, 1926.

28. Although India was the biggest producer, it consumed most of its sugar production domestically; therefore, the Indies became the biggest exporter of sugar in lieu of India (see Bosma, *Sugar Plantation in India and Indonesia*).

29. Alinoeso, "Methods of the White Terror," 3–4.

30. "Uit get Hertogspark" [From Hertogspark], *De Indische Courant*, November 23, 1925.

31. "De Sarekat Hedjo" [The Sarekat Hedjo], *De Sumatra Post*, December 21, 1926.

32. Shiraishi, *Age in Motion*, 312, 314; McVey, *Rise of Indonesian Communism*, 295.

33. Alinoeso, "The Methods of the White Terror," 4–5.

34. Fritjof Tichelman discusses the various regulatory products released as the government's response to the communist movement on chapter 12 of his *Socialisme in Indonesië: Bronnenpublicatie: De Indische Sociaal-Democratische Vereeniging, 1897–1917* [Socialism in Indonesia: Source publication: The Indische social-democratic association, 1897–1917] (Dordrecht, the Netherlands: Foris, 1985), 142–163.

35. On French revolution and the acknowledgement of human rights, including political and civil rights, see Lynn Hunt, ed., *The French Revolution and Human Rights: A Brief Documentary History* (Boston: Bedford/St. Martin's, 1996).

36. *Recht van vereeniging en vergadering* [Right of association and assembly] (Weltevreden, the Netherlands: Landsdrukkerij, 1928), 1. Also see S. Koster and P. Dekker, *Handboek voor politieambtenaren* [Handbook for police officers] (Weltevreden, the Netherlands: Politiegebied te Weltevreden, 1930).

37. Nicholas Tarling, *Nationalism in Southeast Asia: "If the People Are with Us"* (New York: Routledge, 2012), 109.

38. *Recht van vereeniging en vergadering*, 2.

39. Shiraishi, *Age in Motion*, 113.

40. Maters, *Dari perintah halus ke tindakan keras*, 174.

41. This is from the royal decree in Stb. 1919, no. 27, which modified the law Stb. 542 art. 111 in 1915. *Recht van vereeniging en vergadering*, 2. Author's translation.

42. Stb. 1919, no. 27, art. 3, *Recht van vereeniging en vergadering*, 35. Author's translation.

43. *Recht van vereeniging en vergadering*, 17.

44. *Recht van vereeniging en vergadering*, 35. Author's translation.

45. *Recht van vereeniging en vergadering*, 36. Author's translation.

46. *Recht van vereeniging en vergadering*, 37–38. Author's translation.

47. On the reformation of the police institution during the communist movement, see Marieke Bloembergen's works, including *Polisi zaman Hindia belanda: Dari kepedulian dan ketakutan* [Dutch East Indies police: From concern and fear] (Jakarta: Kompas, 2011); "The Dirty Work of Empire: Modern Policing and Public Order in Surabaya, 1911–1919," *Indonesia* 83 (April 2007): 119–50; and "The Perfect Policeman: Colonial Policing, Modernity, and Conscience on Sumatra's West Coast in the Early 1930s," *Indonesia* 91 (April 2011): 165–91.

48. Musso, "Current Information about Indonesia," written in Moscow, April 1, 1929, 5, in "Stukken ingekomen bij het Oost-Secretariaaat van het EKKI betreffende verbanningen" ("Documents received by the Eastern Secretariat of the EKKI regarding bans"), collection ID: ARCH01744, no. 12, Archief Komintern—Partai Komunis Indonesia, IISH, Amsterdam.

49. "Art. 161 bis" [Article 161 bis], *Sinar Hindia*, September 29, 1923.

50. "Fatsal 161 bis dan Tjaboetan Vergadering" [Article 161 bis and meeting withdrawal], *Sinar Hindia*, May 16, 1923. Author's translation.

51. "Derma Delict" [Offense on donation], *Sinar Hindia*, June 6, 1923.

52. Alinoeso, "Methods of the White Terror," 5.

53. Alinoeso, "Methods of the White Terror," 5–6.

54. Sunario, *Het recht van vereeniging en vergadering* [The right of association and meeting] (Leiden, the Netherlands: Perhimpoenan Hakim Indonesia, 1926), 14. Author's translation.

55. Sunario, *Het recht van vereeniging en vergadering*, 14–15.

56. SAM, "Artikel 165 [*sic*] strafwetboek" [Article 165 (*sic*) of the Criminal Code], *Sinar Hindia*, July 16, 1923. Author's translation.

57. Alinoeso, "Methods of the White Terror," 3–4.

58. "Spreekdelikt jang pertama di Indonesia"

59. "Nasib tahoen 1925" [The fate of 1925], *Api*, January 2, 1926.

60. Maters, *Dari perintah halus ke tindakan keras*, 179–80.

61. *Recht van vereeniging en vergadering*, 169.

62. Wiratraman, "Press Freedom, Law and Politics in Indonesia," 48.

63. Wiratraman, "Press Freedom, Law and Politics in Indonesia," 53.

64. Maters, *Dari perintah halus ke tindakan keras*, 209.

65. Maters, *Dari perintah halus ke tindakan keras*, 209n80. Author's translation.

66. Maters, *Dari perintah halus ke tindakan keras*, 212.

67. Maters, *Dari perintah halus ke tindakan keras*, 217.

68. Maters, *Dari perintah halus ke tindakan keras*, 214.

69. Maters, *Dari perintah halus ke tindakan keras*, 225.

70. Musso, "Current Information about Indonesia," 5.

71. Maters, *Dari perintah halus ke tindakan keras*, 229. Author's translation.

72. Maters, *Dari perintah halus ke tindakan keras*, 229.

73. Alinoeso, "Methods of the White Terror," 5.

74. Staatsblad van Nederlandsch-Indie 1931 no. 394 jo. Staatsblad van Nederlandsch-Indie 1932 no. 44. The original title is *Rechtwezen. Drukwerken. Bescherming van de openbare order tegen ongewenschte periodiek verschijnende drukwerken* [Law. Printed matter. Protection of public order against undesirable periodically appearing printed matter].

75. Staatsblad van Nederlandsch-Indie 1931 no. 394. Author's translation.

76. Maters, *Dari perintah halus ke tindakan keras*, 223.

77. Also see Bagus Aries Sugiharta, "The Wild Schools Ordinance in the Dutch East Indies, 1932–1933" (Bachelor's thesis, Leiden University, 2014).

78. Kenji Tsuchiya, *Democracy and Leadership: The Rise of the Taman Siswa Movement in Indonesia* (Honolulu: University of Hawai'i Press, 1987), 152.

79. "Onderwijs ordonnantie dengan praktijknja D.l.l." [The Education Ordinance and its practice, etc.], *Sinar Hindia,* July 30, 1924.

80. *Menentang wilde scholen ordonnantie* [Resisting wild school ordinance] (Jakarta: Perhimpunan Pelajar-Pelajar Indonesia, 1933), 5.

81. Sugiharta, "Wild Schools Ordinance in the Dutch East Indies"; and Ewout Frankema, "Why Was the Dutch Legacy So Poor? Educational Development in the Netherlands Indies, 1871–1942" *Masyarakat Indonesia* 39, no. 2 (December 2013), 322.

82. *Menentang wilde scholen ordonnantie,* 5. Author's translation.

83. *Menentang wilde scholen ordonnantie,* 4. Author's translation.

84. *Menentang wilde scholen ordonnantie,* 3–4.

85. *Menentang wilde scholen ordonnantie,* 7.

86. *Recht van vereeniging en vergadering,* 17.

87. *Recht van vereeniging en vergadering,* 8. Emphasis added.

88. Tsuchiya, *Democracy and Leadership,* 151.

89. Krishna Sen and David T. Hill, *Media, Culture and Politics in Indonesia* (Melbourne: Oxford University Press, 2000).

90. On regulations of media historically in other parts of the world, some were invented to protect freedom of expressions and some were to limit it: Robert Darnton and Daniel Roche, eds., *Revolution in Print: The Press in France, 1775–1800* (Berkeley: University of California Press, 1989); and Robert Darnton, *The Literary Underground of the Old Regime* (Cambridge, MA: Harvard University Press, 1982).

91. Maters, *Dari perintah halus ke tindakan keras,* 239. See PPO in Harry Poeze, *Politiek-politioneele overzichten van Nederlandsch-Indië. Een bronnenpublikatie. Deel I 1927–1928* [Political-police overviews of the Dutch East Indies. A source publication. Part I 1927–1928] (Den Haag, the Netherlands: KITLV, 1982); *Politiek-politioneele overzichten van Nederlandsch-Indië. Een bronnenpublikatie. Deel II 1929–1930* [Political-police overviews of the Dutch East Indies. A source publication. Part II 1929–1930] (Den Haag, the Netherlands: KITLV, 1983); *Politiek-politioneele overzichten van Nederlandsch-Indië. Een bronnenpublikatie. Deel III 1931–1934* [Political-police overviews of the Dutch East Indies. A source publication. Part III 1931–1934] (Dordrecht/Providence, the Netherlands: KITLV, 1987); and *Politiek-politioneele overzichten van Nederlandsch-Indië. Een bronnenpublikatie. Deel IV 1935–1941* [Political-police overviews of the Dutch East Indies. A source publication. Part IV 1935–1941] (Den Haag, the Netherlands: KITLV, 1994).

92. Maters, *Dari perintah halus ke tindakan keras.*

93. Rudolf Mrázek, *The Complete Lives of Camp People: Colonialism, Fascism, Concentrated Modernity* (Durham, NC: Duke University Press, 2020), 143–208; and Shiraishi, "Phantom World of Digoel," 108–9.

7. THE OTHER LABOR OF CLANDESTINE SAILORS

1. Timorman, "Korban-korban PARI" [The victims from PARI], written in August 1933, in "Deels gecodeerde brieven van Indonesische communisten aan 'Anton'" ("Partly coded letters from Indonesian communists to 'Anton'"), collection ID: ARCH01744, no. 29, Archief Komintern—Partai Komunis Indonesia, International Institute of Social History (IISH), Amsterdam. Most of PARI history in this document was

republished in Djamaloedin Tamin's memoir using the spelling Djamaluddin Tamim, "Sedjarah PKI Djilid III" (unpublished manuscript, 1957).

2. See Kris Alexanderson, *Subversive Seas: Anticolonial Networks Across the Twentieth-Century Dutch Empire* (Cambridge: Cambridge University Press, 2019); Peter Linebaugh and Marcus Rediker, "The Many-Headed Hydra: Sailors, Slaves, and the Atlantic Working Class in the Eighteenth Century," *Journal of Historical Sociology* 3, no. 3 (September 1990), 230; Matthias van Rossum, *Hand aan hand (Blank en bruin): Solidariteit en de werking van globalisering, etniciteit en klasse onder zeelieden op de Nederlandse koopvaardij, 1900–1945* [Hand in hand (white and brown): Solidarity and the effect of globalization, ethnicity and class among sailors in the Dutch merchant navy, 1900–1945] (Amsterdam: Aksant, 2009); also on the 1933 mutiny by sailors on board of De Zeven Proviciën, see Jaap R. Bruijn and Els van Eyck van Heslinga, "Mutiny: Rebellion on the Ships of the Dutch East India Company," *The Great Circle* 4, no. 1 (April 1982), 1–9.

3. See photo in Harry A. Poeze, Cees van Dijk, and Inge van der Meulen, *Di Negeri Penjajah: Orang Indonesia di Negeri Belanda 1600–1950* (Jakarta: KPG and KITLV Jakarta, 2014), 184; from "Affiche met aankondiging welkomstvergadering ter begroeting van de Indonesier Semaoen" ("Poster announcing a welcome meeting to greet the Indonesian Semaoen"), collection ID: ARCH00321, no. 45–2, Collectie CPH, IISH, Amsterdam. Author's translation.

4. Sarikat Pegawei Laoet Indonesia, SPLI, no. 12041, *Rapporten Centrale Inlichtingendienst, 1919-1940*, historici.nl, now in Huygens Instituut, Amsterdam, https://resources. huygens.knaw.nl/rapportencentraleinlichtingendienst/data/IndexResultaten/IndexVenst erResultaat?persoon=Semaoen+%28alias+Soleiman%29. Also see Semaun, *Skets sedjarah Pak Matosin* (Surabaja, Indonesia: N. V. Matang, 1962), 46.

5. Quoted from M. A., "Die arbeiterbewegung in Indonesien," in Ruth McVey, *The Rise of Indonesian Communism* (Ithaca, NY: Cornell University Press, 1965), 442n78.

6. Agustinus Supriyono, *Buruh pelabuhan Semarang: Pemogokan-pemogokan pada zaman kolonial belanda, revolusi dan republik, 1900–1965* (PhD diss., Vrije Universiteit Amsterdam, 2008).

7. Quoted from M. A., "Die arbeiterbewegung in Indonesien," 442n78.

8. McVey, *Rise of Indonesian Communism*, 276–77.

9. See "Receipt Mei Juni [*sic*] and July 1924" and a letter to the Eastern Department of Comintern and Profintern dated Moskow November 15, 1924, from 'Bericht des Genossen Semaoen über die KP Indonesien' en bij Comintern en Profintern ingekomen brieven en rapporten van Tan Malakka en H. Maring (Sneevliet). ('Report of Comrade Semaoen on the Communist Party of Indonesia' and letters and reports from Tan Malakka and H. Maring (Sneevliet) received by the Comintern and Profintern.). Archief Komintern—Partai Komunis Indonesia, collection ID: ARCH01744, no. 10, IISH, Amsterdam.

10. Onimaru Takeshi, "Living 'Underground' in Shanghai: Noulens and the Shanghai Comintern Network," in *Traveling Nation-Makers: Transnational Flows and Movements in the Making of Modern Southeast Asia*, ed. Caroline Hau and Kasian Tejapira (Singapore: NUS Press, 2011), 100.

11. Alexanderson, *Subversive Seas*, 179.

12. Onimaru Takeshi, "Shanghai Connection: The Construction and Collapse of the Comintern Network in East and Southeast Asia," *Southeast Asian Studies* 5, no. 1 (2016): 116.

13. Takeshi, "Shanghai Connection," 127.

14. Audrey Kahin, *Rebellion to Integration: West Sumatra and the Indonesian Polity, 1926–1998* (Amsterdam: Amsterdam University Press, 1999), 62.

15. Timorman, "Korban-korban PARI" [The victims from PARI], 5–7.

16. Timorman, "Korban-korban PARI," 5–7.

17. Timorman, "Korban-korban PARI."

18. John Riddell, *Toward the United Front: Proceedings of the Fourth Congress of the Communist International, 1922* (Chicago: Haymarket Books, 2012), 261–65. Vladimir Lenin often revised his statement upon hearing the opinions of leaders from the South including his recognition of Palestinian struggles. Ran Greenstein, *Zionism and Its Discontents: A Century of Radical Dissent in Israel/Palestine* (London: Pluto Press, 2014), 50–103.

19. Audrey Kahin, *Rebellion to Integration*, 62.

20. Recent work that provides findings on PKI activities beyond the 1926 to 1927 revolts includes Kankan Xie, "Estranged Comrades: Global Networks of Indonesian Communism, 1926–1932" (PhD Diss., University of California, Berkeley, 2018).

21. "Proposals for Exchange of Information on Communism Between India, the Dutch East Indies, and French Indo-China," April 1925–December 1926, file number: L/P&J/12/249; file 795/1925, British Library, London. Also see Anne L. Foster, "Secret Police Cooperation and the Roots of Anti-Communism in Interwar Southeast Asia," in *Journal of American-East Asian Relations* 4, no. 4 (1995), 331–50.

22. Alexanderson, *Subversive Seas*, 170.

23. Alexanderson, *Subversive Seas*, 172. An example of a cooperation between colonial states in Southeast Asia could be seen in a letter of agreement in the "Proposals for Exchange of Information."

24. Alexanderson, *Subversive Seas*, 184.

25. Alexanderson, *Subversive Seas*, 208.

26. Timorman, "Korban-Korban PARI." Also see Kankan Xie, "The Netherlands East Indies 1926 Communist Revolt Revisited: New Discoveries from Singapore's Digital Newspaper Archives," in *Chapters on Asia: Selected Papers from the Lee Kong Chian Research Fellowship (2014–2016)* (Singapore: National Library Board, 2018), 267–94.

27. *Malayan Bulletin of Political Intelligence*, June 1927, courtesy of Audrey Kahin's personal collection.

28. *Malayan Bulletin of Political Intelligence*, July 1927.

29. Timorman, "Korban-Korban PARI."

30. Timorman, "Korban-Korban PARI."

31. The fingerprint collection was a relatively recent technique used to record the identity of crew members. Around 1918, port cities set up a company security service to be in charge of a dactyloscopy card archive that kept the fingerprints and other data of the many thousands of crew members "in order to be able to carry out checks when recruiting these people where names and photographs cannot be relied on." See Joseph Norbert Frans Marie à Campo, *Engines of Empire: Steamshipping and State Formation in Colonial Indonesia* (Hilversum, the Netherlands: Verloren, 2002), 455n301.

32. Audrey Kahin, *Rebellion to Integration*, 47–48.

33. Audrey Kahin, *Rebellion to Integration*, 54.

34. Timorman, "Korban-Korban PARI" ["The victims from PARI"], 7.

35. Tamim, "Sedjarah PKI Djilid III," 57.

36. See Harry A. Poeze, *Tan Malaka: Strijder voor Indonesië's vrijheid: Levensloop van 1897 tot 1945* [Tan Malaka: Fighter for Indonesia's freedom, life history from 1897 to 1945] ('s-Gravenhage: Nijhoff, 1976); Rudolf Mrázek, "Tan Malaka: A Political Personality's Structure of Experience," *Indonesia* 14 (October, 1972): 1–48.

37. James R. Rush, *Opium to Java: Revenue Farming and Chinese Enterprise in Colonial Indonesia 1860–1910* (Jakarta: Equinox Publishing, 2007 [1990]), 230–31.

38. In an interview with Ruth McVey in 1959, Semaoen recalled that the first government action against this smuggling activity was a raid on the incoming passenger ship Insulinde in Java. Police confiscated a suitcase full of letters and publications that a cabin boy acted as the "consul" for the SPLI on that ship had brought in. Among the publications

were *Pandoe Merah* and other communist materials. See McVey, *Rise of Indonesian Communism,* 452n181.

39. Sarikat Pegawei Laoet Indonesia, SPLI, no. 12041.

40. Sarikat Pegawei Laoet Indonesia, SPLI, no. 12041.

41. Semaun, *Skets sedjarah Pak Matosin*, 40.

42. Semaun, *Skets sedjarah Pak Matosin*.

43. Letters to the representative of the Comintern dated September 23, 1924, and October 15, 1924, collection ID: ARCH01744, no. 10, Archief Komintern—Partai Komunis Indonesia, IISH, Amsterdam.

44. "Bericht des Genossen Semaoen über die KP Indonesien," Archief Komintern—Partai Komunis Indonesia.

45. Sarikat Pegawei Laoet Indonesia, SPLI, no. 12041.

46. "Affiche 'Weg met Welter, Ruys en Colijn! Staakt tegen loonsverlagingen. Kiest Communisten'" ("Poster 'Away with Welter, Ruys, and Colijn! Strike against wage cuts. Vote Communists'"), collection ID: ARCH00321, no. 45–2, Collectie CPH, IISH, Amsterdam. Author's translation.

47. Indonesians continued to involve in global anti-imperialist struggles. See Fredrik Petersson, "Hub of the Anti-Imperialist Movement: The League against Imperialism and Berlin, 1927–1933," *Interventions: International Journal of Postcolonial Studies* (2013): 1–23; Fredrik Petersson, *Willi Münzenberg, the League against Imperialism, and the Comintern, 1925–1933* (Lewiston, NY: Edwin Mellen Press, 2014); Klaas Stutje, *Campaigning in Europe for a Free Indonesia: Indonesian Nationalists and the Worldwide Anticolonial Movement, 1917–1931* (Copenhagen: NIAS Press, 2019).

48. Semaun, *Skets sedjarah Pak Matosin*, 50.

49. Sarikat Pegawei Laoet Indonesia, SPLI, no. 12041.

50. Tan Malaka, *From Jail to Jail 1*, trans. Helen Jarvis (Athens: Ohio University Center for International Studies, 1991), 111.

51. Sarikat Pegawei Laoet Indonesia, SPLI, no. 12041. This conference followed the famous Hong Kong-Canton Seamen's Strike of 1922. Ming K. Chan, "Labor in Modern and Contemporary China," *International Labor and Working-Class History* 11 (May 1977): 13–18.

52. On KUTV: Masha Kirasirova, "The 'East' as a Category of Bolshevik Ideology and Comintern Administration: The Arab Section of the Communist University of the Toilers of the East," *Kritika: Explorations in Russian and Eurasian History* 18, no. 1 (Winter 2017): 7–34. On Iranian students at the KUTV, see: Lana Ravandi-Fadai, "'Red Mecca'—The Communist University for Laborers of the East (KUTV): Iranian Scholars and Students in Moscow in the 1920s and 1930s," *Iranian Studies* 48, no. 5 (2015): 713–27. On students from Africa, see: Irina Filatova, "Introduction or Scholarship? Education of Africans at the Communist University of the Toilers of the East in the Soviet Union, 1923–37," *Paedagogica Historica* 35, no. 1 (1999): 41–66. For a more complete list of bibliography on KUTV in multiple languages, see Kirasirova, "'East' as a Category," 10n12–13.

53. Sometime in 1927, Waworoentoe and Wentoek left for the Indies to propagate communism, but they were arrested as soon as they arrived there. The news of their arrest spread out to the Netherlands, including in *New Rotterdamsche Courant* on March 18, 1929, which contained a cable from Indonesia detailing their journey to Moscow as well as their communist missions for returning to the Indies after the revolt, in "Stukken ingekomen bij het Oost-Secretariaat van het EKKI betreffende verbaningen" ("Documents received by the Eastern Secretariat of the EKKI concerning banishments"), collection ID: ARCH01744, no. 12, Archief Komintern—Partai Komunis Indonesia, IISH, Amsterdam. Though similar in name, the sailor Waworoentoe was not related to the well-known Volksraad member.

54. Alexanderson, *Subversive Seas*, 257–59, includes "Appendix: Testimony from Communist Informant Kamu [Kamoe], 26 January 1928." While the title clearly states that the testimony comes from Kamoe, whose first name is Daniel, in the testimony Kamoe pretends that he is not Daniel Kamoe, saying, "I did not speak to Semaoen by myself; during the conversations there were a few other seamen present named Clement Wentoek, Johannes Wawoeroentoe and someone named Daniel, whose last name I do not know" (257). This indicates that the informant is not Kamoe and might be wrongly taken as Kamoe. He also claims to be sent back to the Indies with Wentoek, but Wentoek returned with Waworoentoe (see "Stukken ingekomen bij het Oost-Secretariaat van het EKKI betreffende verbaningen" [Documents received by the Eastern Secretariat of the EKKI concerning banishments], collection ID: ARCH01744, no. 12, Archief Komintern—Partai Komunis Indonesia, IISH, Amsterdam). It seems that the informant intentionally mixed up the details to confuse officials. Aside from minor discrepancies, however, his story about the travel to Russia and the KUTV situation appears to be sound.

55. Alexanderson, *Subversive Seas*, 258.

56. "Materials and reports of the academic part for the 1925-1926 academic year," box: F. 532, op. 8, case: 78; "Minutes of the sessions of the preparatory course of the KUTW for 1925/1926," box: F. 532, op. 8, case: 89; "Curriculum, calendar plans, programs for the 1926/1927 school year," box: F. 532, op. 8, case: 135, Russian State Archive of Socio-Political History (RGASPI), Moscow.

57. "Student lists and Dengli theses on the Chinese Question," box: F. 532, op. 1, case: 388, RGASPI, Moscow. While in Russia, Soemantry married a Russian woman Dengli Anna Ilyinicna, and together they had a child named Mirinou (name is unclear) Dengli. See "Personal file Dengli Anna Illyinicna," box: F. 495, op. 214, case: 42, RGASPI, Moscow.

58. Quotes are from "Personal file Mina Khasa," box: F. 495, op. 214, case: 64a, RGASPI, Moscow. For sources on Oesman (Banka), see "Personal file Banka," box: F. 495, op. 214, case: 62; on Soemantry (Dingli), see "Personal file of Dengly Sinavi," box: F. 495, op. 214, case: 63; on Mohamad Saleh (Mulia), see "Personal file Mulya (Mulia)," box: F. 495, op. 214, case: 64; on Waworoentoe (Minahasa), see "Personal file Mina Khasa," box: F. 495, op. 214, case: 64a; on Clemens Wentoek (Passi), see "Passi Personal file," box: F. 495, op. 214, case: 65; on Daniel Kamoe (Celebes), see "Private file Slebis," box: F. 495, op. 214, case: 66, RGASPI, Moscow.

59. "Deels gecodeerde brieven van Indonesische communisten aan 'Anton'," Archief Komintern—Partai Komunis Indonesia.

60. Timorman, "Korban-korban PARI."

61. One could argue that these sailors do not fit in the traditional concept of labor because they do not produce commodities in factories described in Marx's *Capital: A Critique of Political Economy Vol. 1* (London: Penguin Books, in association with New Left Review, 1990). I argue that sailors are important parts of production network because, by producing mobility and transporting goods and people, they turn mere products into commodity. In other words, without transport, products will remain to be products in factories. They become commodity as they are circulated and distributed to reach the market for consumption. Therefore, port workers and sailors are indispensable part in the (global) chain of commodity production.

62. Karl Marx, "Economic and Philosophic Manuscripts of 1844," in *The Marx-Engels reader,* ed. Robert C. Tucker (New York: Norton, 1978), 73. Italics in original.

63. Marx, "Economic and Philosophic Manuscripts of 1844," 74.

64. Marx, "Economic and Philosophic Manuscripts of 1844," 75.

65. Marx, "Economic and Philosophic Manuscripts of 1844," 75.

66. Michel Foucault alludes to this dialectic in his discussion of heterotopia. See Michel Foucault, "Of Other Spaces: Utopias and Heteropias," in *Rethinking Architecture: A Reader in Cultural Theory*, ed. Neil Leach (New York: Routledge, 1997), 330–36.

67. Cf. Nick Nesbitt, *Universal Emancipation: The Haitian Revolution and the Radical Enlightenment* (Charlottesville: University of Virginia Press, 2008).

68. Marwan Kraidy also explores resistance work from the perspective of labor by using the term "revolutionary creative labor" in "Revolutionary Creative Labor," in *Precarious Creativity: Global Media, Local Labor*, ed. Michael Curtin and Kevin Sanso (Oakland: University of California Press, 2016), 231–40.

69. Timorman, "Korban-korban PARI."

70. E. P. Thompson, "Time, Work-Discipline, and Industrial Capitalism," *Past & Present*, no. 38 (December 1967): 80.

71. Thompson, "Time, Work-Discipline, and Industrial Capitalism," 93–94.

72. Timorman, "Korban-korban PARI," 14.

73. Timorman, "Korban-korban PARI."

74. Timorman, "Korban-korban PARI," 15–16.

75. Timorman, "Korban-korban PARI."

76. Timorman, "Korban-korban PARI," 27.

77. Timorman, "Korban-korban PARI," 27.

78. Timorman, "Korban-korban PARI," 27.

79. Timorman, "Korban-korban PARI," 29.

80. Cf. Jodi Dean, *Crowds and Party* (London: Verso, 2018).

81. Campo, *Engines of Empire*, 453.

82. Campo, *Engines of Empire*, 453.

83. Campo, *Engines of Empire*, 454.

84. Jan Slauerhoff, *Kubur terhormat bagi pelaut* (Jakarta: Pustaka Jaya, 1977 [1936]).

85. Campo, *Engines of Empire*, 53.

86. Campo, *Engines of Empire*, 454–55.

87. Linebaugh and Rediker, "Many-Headed Hydra," 244.

88. Semaun, *Skets sedjarah Pak Matosin*, 22–24.

89. Semaun, *Skets sedjarah Pak Matosin*, 34.

90. A letter to EECI Eastern Department Moscow dated Berlin, September 11, 1924, in "'Bericht des Genossen Semaoen über die KP Indonesien,'"Archief Komintern—Partai Komunis Indonesia.

91. Timorman, "Korban-korban PARI," 68.

92. Inspired by a famous British 1905 novel *The Scarlet Pimpernel*, Matu Mona wrote its Indonesian version, a five-volume *Patjar Merah Indonesia* (*The Scarlet Pimpernel of Indonesia*) published between 1938 and 1940 in Medan, East Sumatra. In this novel, Tan Malaka, as the Scarlet Pimpernel of Indonesia, is depicted as constantly facing danger and enemies traversing different countries—China, Russia, the Philippines, Thailand, Germany, Iran, and Palestine. Malaka in the novel seems to be curiously always able to maintain the freedom to escape. Noriaki Oshikawa, "Patjar Merah Indonesia and Tan Malaka: A Popular Novel and a Revolutionary Legend," in *Reading Southeast Asia: Translation of Contemporary Japanese Scholarship on Southeast Asia*, ed. Takashi Shiraishi, vol. 1 (Ithaca, NY: Cornell University Press, 1990), 32–33. The novel clearly mixes facts, fictions, and myths about Tan Malaka during his two decades of exile from the Indies since 1922.

93. Oshikawa, "'Patjar Merah Indonesia," 33n56.

94. Linebaugh and Rediker, "Many-Headed Hydra," 228. Elsewhere, Todd Gitlin, in *The Whole World Is Watching: Mass Media in the Making and Unmaking of the New Left* (Berkeley: University of California Press, 2003), explains how mainstream media, government, and people from outside the movement often turned movement leaders into

celebrities, estranging them from their base, which, in turn, undermined the importance of collective work that ordinary people contributed in a movement.

CODA

1. It is very curious that despite her main role in the movement, Djoeinah escaped the limelight of the mainstream media at the time. Her arrest in Digoel was only confirmed in the long list of Digoel internees as the wife of the Vereniging van Spoor-en Tramwegpersoneel leader Prapto, who was arrested around the same time as Moenasiah. See "Gubernur Jawa Barat dan Sumatera Utara kepada Hakim Tinggi Hindia Belanda: Surat-surat bulan Maret-Juli 1927 tentang pengiriman narapidana ke Tanah Merah. NB. Beserta daftar nama-nama narapidana tersebut" ("Governor of West Java and North Sumatra to the High Judge of the Dutch East Indies: Letters from March-July 1927 regarding the sending of prisoners to Tanah Merah. PS. Along with a list of the names of the prisoners"), no. 200, Inventaris arsip Boven Digoel (Boven Digoel archive inventory), National Archives of Indonesia (ANRI), Jakarta. In other words, she was not officially arrested as a political prisoner; she was considered an accompanying family member to join Prapto. *Woro* Ati was already arrested in Garut due to speech offense (chapter 6) before she was exiled to Digoel, however. See "Naar Boven Digoel de nieuwe lijst" [To Boven Digoel the new list], *De Indische Courant*, February 8, 1927; "Geïnterneerden" [Internees], *De Locomotief*, June 13, 1927; "Voor Boven-Digoel Geïnterneerden" [For Boven-Digoel internees], *De Locomotief*, February 21, 1927; and "Omverwerping van het Ned. Gezag in Oost-Indie" [The Overthrow of the Netherlands Authority in the East Indies], *Onze Courant*, 12 July 1929. Many of these people were released from Digoel by the 1930s. After her exile in Digoel, *woro* Moenasiah remained an important figure in the Communist Party of Indonesia (PKI). See her name listed in the sixth Communist Party of Indonesia's national congress in 1959, "Pidato pembukaan kongres nasional Ke-VI PKI Jang Diutjapkan Pada Tgl 7 September 1959" [The opening speech of the Sixth National Congress of the PKI delivered on September 7, 1959], no. 309, KOTI (Komando Operasi Tertinggi), ANRI, Jakarta.

2. See especially Takashi Shiraishi's discussion of the *pergerakan* in *An Age in Motion: Popular Radicalism in Java, 1912–1926* (Ithaca, NY: Cornell University Press, 1990), 339–42. Other work includes Adrian Vickers, *A History of Modern Indonesia* (Cambridge: Cambridge University Press, 2013 [2005]); M. C. Ricklefs, *A History of Modern Indonesia since c. 1200* (Stanford, CA: Stanford University Press, 2001 [1991]), 318–19; Harry A. Poeze, "Early Indonesian Emancipation: Abdul Rivai, van Heutsz and the *Bintang Hindia*," *Bijdragen tot de Taal-, Land- en Volkenkunde* 145 (1989): 87–106; and Benedict Anderson, *Language and Power: Exploring Political Cultures in Indonesia* (Ithaca, NY: Cornell University Press, 1990).

3. Resonating with this, Harry J. Benda and Ruth T. McVey argue that at this time the movement was still proto-nationalist and not yet nationalist in character. Harry J. Benda and Ruth T. McVey, *The Communist Uprisings of 1926–1927 in Indonesia: Key Documents* (Ithaca, NY: Modern Indonesia Project, Cornell University, 1960), xi.

4. Clifford Siskin, "Mediated Enlightenment: The System of the World," in *This Is Enlightenment*, ed. Clifford Siskin and William Warner (Chicago: University of Chicago, 2010), 164.

5. See Jonathan Israel, *Radical Enlightenment* (Oxford: Oxford University Press, 2001); and Margaret Jacob, *The Radical Enlightenment: Pantheists, Freemasons and Republicans* (London: George Allen and Unwin, 1981). Sankar Muthu describes this radical tradition as Enlightenment anti-imperialism. Sankar Muthu, *Enlightenment against Empire* (Princeton, NJ: Princeton University Press, 2009).

6. Carla Hesse, *The Other Enlightenment: How French Women Became Modern* (Princeton, NJ: Princeton University Press, 2001); Sebastian Conrad, "Enlightenment in Global History: A Historiographical Critique," *American Historical Review* (October 2012): 999–1027; Jose S. Arcilla, "The Enlightenment and the Philippine Revolution," *Philippine Studies* 39, no. 3 (Third Quarter 1991), 358–73 (for further literature on the Philippines see note in chapter 2, note 27); and Vera Schwarcz, *The Chinese Enlightenment: Intellectuals and the Legacy of the May Fourth Movement of 1919* (Berkeley: University of California Press, 1986).

7. Terms such as "postcolonial Enlightenment," "provincializing Europe," "alternative modernities," and "multiple modernities" are some examples of this attempt to de-Westernize the history of modernity and Enlightenment from the lens of difference. See Daniel Carey and Lynn M. Festa, *The Postcolonial Enlightenment: Eighteenth-Century Colonialism and Postcolonial Theory* (Oxford: Oxford University Press, 2015); Dipesh Chakrabarty, *Provincializing Europe: Postcolonial Thought and Historical Difference* (Princeton, NJ: Princeton University Press, 2008); S. N. Eisenstadt, "Multiple Modernities," *Daedalus* 129, no. 1 (Winter 2000): 1–29; and Nilüfer Göle, "Snapshots of Islamic Modernities," *Daedalus* 129, no. 1 (Winter 2000), 91–118. These works essentially operate within a particular universalism, that is otherness. In my view, in thinking about commonalities, accounting to differences is important without dismissing the shared universal. For a thorough conceptualization of this account that takes both universalism and difference as equally important, see Vivek Chibber, *Postcolonial Theory and the Specter of Capital* (London: Verso, 2013).

8. On the problem of the poststructuralist turn to difference: Marnia Lazreg, *Foucault's Orient: The Conundrum of Cultural Difference, From Tunisia to Japan* (New York: Berghahn Books, 2017); and Peter Dews, *Logics of Disintegration: Post-Structuralist Thought and the Claims of Critical Theory* (London: Verso, 2007).

9. Cf. Raymond Williams, *The Long Revolution* (Cardigan, UK: Parthian, 2011).

10. Conrad, "Enlightenment in Global History," 1027.

11. Jürgen Habermas, *The Structural Transformation of the Public Sphere: An Inquiry into a Category of Bourgeois Society* (Cambridge, MA: MIT Press, 1989). On the plebeian public sphere, see Geoff Eley, "Nations, Publics, and Political Cultures: Placing Habermas in the Nineteenth Century" in *Habermas and the Public Sphere*, ed. Craig Calhoun (Cambridge, MA: MIT Press, 1992), 289–339. On Bengali *adda*, see Dipesh Chakrabarty, *Provincializing Europe: Postcolonial Thought and Historical Difference* (Princeton, NJ: Princeton University Press, 2008), 180–213. On the revolutionary public sphere, see Rianne Subijanto, "Communist *Openbare Vergaderingen* and an Indonesian Revolutionary Public Sphere," in *The Global Impacts of Russia's Great War and Revolution Book 2: The Wider Arc of Revolution, Part 2*, ed. Choi Chatterjee, Steven G. Marks, Mary Neuberger, and Steven Sabol (Bloomington: Slavica Publishers, 2019), 277–97. Also see David Zaret, *Origins of Democratic Culture: Printing Petitions and the Public Sphere in Early-Modern England* (Princeton NJ: Princeton University Press, 2000).

12. Karl Marx, "Manifesto of the Communist Party," 1848, https://www.marxists.org/archive/marx/works/1848/communist-manifesto/.

13. Todd Gitlin, "Why 'the Enlightenment Project' Is Necessary and Unending," *Tablet*, February 17, 2015, https://www.tabletmag.com/sections/news/articles/the-enlightenment-project.

Archival Sources

Archives

Arsip Nasional Republik Indonesia (ANRI), Jakarta, Indonesia

    Inventaris Arsip Boven Digoel

    Inventaris Arsip Post-, Telegraaf- en Telefoondienst 1817–1950

    Komando Operasi Tertinggi (KOTI)

British Library, London, UK

historici.nl (now in huygens.knaw.nl, Huygens Instituut, Amsterdam, The Netherlands)

    Rapporten Centrale Inlichtingendienst, 1919–40

International Institute of Social History (IISH), Amsterdam, The Netherlands

    Archief Komintern—Partai Komunis Indonesia

    Collectie CPH

The Royal Netherlands Institute of Southeast Asian and Caribbean Studies (KITLV) Digital Image Library, Leiden, The Netherlands

The Carl A. Kroch Library, Cornell University, Ithaca, NY, USA

Nationaal Archief (NA), the Hague, the Netherlands

    Collectie Documentatiebureau voor Overzees Recht, 1894–1963

Russian State Archive of Socio-Political History (RGASPI), Moscow, Russia

    Executive Committee of the Comintern (ECCI) (1919–43)

    Communist University of the Toilers of the East (KUTV) (1921–38); Research Institute of National and Colonial Problems (NII NKP) (1936–39)

Indonesian Newspapers Consulted

Api

Djago! Djago!

Doenia achirat

Doenia bergerak

Islam bergerak

Medan bergerak

Medan Moeslimin

Njala
Pandoe merah
Pemandangan Islam
Sinar Hindia
Soeara tambang

Index

www.ingramcontent.com/pod-product-compliance
Lightning Source LLC
Chambersburg PA
CBHW031602110425
24987CB00028B/286